The WTO and Environment-Related International Trade Disputes

Biosecurity and Ecosystem Services Risks

Other World Scientific Titles by the Author

Applications of Principal-Agent Theory to Agricultural Land Use Policy: Lessons from the European Union
by Robert W. Fraser
ISBN: 978-1-78326-675-3

Plant Biosecurity Policy Evaluation: The Economic Impacts of Pests and Diseases
by David C. Cook, Robert W. Fraser and Andrew Wilby
ISBN: 978-1-78634-215-7

The WTO and Environment-Related International Trade Disputes

Biosecurity and Ecosystem Services Risks

Robert W. Fraser
University of Kent, UK

David C. Cook
University of Western Australia, Australia

Janet Haddock-Fraser
Manchester Metropolitan University, UK

World Scientific

NEW JERSEY • LONDON • SINGAPORE • BEIJING • SHANGHAI • HONG KONG • TAIPEI • CHENNAI • TOKYO

Published by

World Scientific Publishing Europe Ltd.
57 Shelton Street, Covent Garden, London WC2H 9HE
Head office: 5 Toh Tuck Link, Singapore 596224
USA office: 27 Warren Street, Suite 401-402, Hackensack, NJ 07601

Library of Congress Cataloging-in-Publication Data
Names: Fraser, Robert W., author. | Cook, David (David C.) (Agricultural economist), author. | Haddock-Fraser, Janet, author.
Title: The WTO and environment-related international trade disputes : biosecurity and ecosystem services risks / Robert W. Fraser, David C. Cook, Janet Haddock-Fraser.
Other titles: World Trade Organization and environment-related international trade disputes
Description: [Hackensack, New Jersey] : World Scientific, 2019. | Includes bibliographical references and index.
Identifiers: LCCN 2019034410 | ISBN 9781786347770 (hardcover)
Subjects: LCSH: Foreign trade regulation. | World Trade Organization. | Environmental risk assessment--Law and legislation. | Biosecurity. | Ecosystem health--Risk assessment.
Classification: LCC K3943 .F73 2019 | DDC 382/.92--dc23
LC record available at https://lccn.loc.gov/2019034410

British Library Cataloguing-in-Publication Data
A catalogue record for this book is available from the British Library.

For any available supplementary material, please visit
https://www.worldscientific.com/worldscibooks/10.1142/Q0231#t=suppl

Desk Editors: Anthony Alexander/Jennifer Brough/Shi Ying Koe

Typeset by Stallion Press
Email: enquiries@stallionpress.com

Foreword

Widespread public policy concern about the natural environment began only half a century ago. In his 1971 Wiksell Lectures in Stockholm, William Baumol was one of the first economists to analyse international spillovers associated with environmental protection policies. Concern at that time was focused mostly on strict industrial pollution standards harming firms and workers in rich countries, for which they demanded compensation in the form of subsidies or protection from imports from countries with lower or no such standards. This was also the subject of the first of the GATT Secretariat's *GATT Studies in International Trade* (1971). Following a lull brought on by the economic disorders of the decade following the first oil price shock in 1973, a second and more-intense wave in public interest in environmental matters arose. It was more widespread, embracing the primary and services sectors in addition to manufacturing; and global issues such as ozone depletion and climate change also came into focus. The GATT Secretariat's Annual Report of 1991 (*International Trade 1990/91*) and an edited volume of its background papers (K. Anderson and R. Blackhurst (eds.), *The Greening of World Trade Issues*, University of Michigan Press, 1992) explored what the linking of trade and the environment by policy influencers might mean for GATT Contracting Parties. Since then, the GATT's Uruguay Round of multilateral trade negotiations brought into being, in January 1995, the World Trade Organization (WTO) and its Dispute Settlement Body (DSB). This led environmental groups to focus even more on what roles trade — and the DSB — might play in protecting the environment nationally and globally. While that perspective did not always sit comfortably with the traditional view of liberal traders that opening markets was good for global welfare, the world's pre-eminent international trade lawyer at the time pointed out that environmentalists are but one of numerous groups that are not unequivocally happy about increasing international economic interdependence (J. Jackson, "Global Economics and International Economic Law", *Journal of International Economic Law* 1(1): 1–23, 1998).

This book is about one aspect of such concerns, namely, the role of the WTO's DSB in managing risks associated with biosecurity and ecosystem services. The issue is tackled not only conceptually but also by examining actual DSU case studies. The latter confirm that while some decisions by the DSB are likely to have enhanced multilateral social, economic and environmental welfare, others are likely to have worsened it. The authors call for greater use by WTO dispute settlement panels of information to assess how a decision could affect producer interests, consumer welfare and the environment. A methodology is provided for making more-comprehensive assessments of this sort. The opportunity is now there for the WTO membership to give this consideration as and when the DSB rules and procedures are next reviewed.

Kym Anderson
University of Adelaide
and
Australian National University

About the Authors

Robert W. Fraser is an Emeritus Professor of Agricultural Economics. He completed a first degree in Economics at Adelaide University before gaining a Rhodes Scholarship to study MPhil and DPhil in Economics at Oxford University from 1978. His first appointment was as an Assistant Professor of Economics at the University of Virginia, USA in 1981. A year later, he joined the University of Western Australia initially as a Lecturer in Economics, becoming Professor of Agricultural and Resource Economics early in 1999. In 2000, he was appointed as the Professor of Agricultural Economics at Imperial College, and in 2006 he joined the University of Kent in the same capacity. He has an international research reputation as a policy economist, specialising in both agri-environmental and invasive species policy design and evaluation. In this context, since moving to the UK in 2000 he has participated in a range of DEFRA and other funded research projects. In addition, he was commissioned in 2006 by the OECD to prepare a report on "Information Deficiencies in Agri-Environmental Policies", which was then presented as the keynote paper to an OECD Workshop on this topic. He was the President of the Agricultural Economics Society (AES) and was both the President and Distinguished Fellow of the Australian Agricultural and Resource Economics Society (AARES) in the past. He is also a member of the editorial board of the *Journal of Agricultural Economics*. Since 2012, he has been a member of DEFRA's Economic Advisory Panel.

David C. Cook is Senior Economist at the Department of Primary Industries and Regional Development (DPIRD) and Adjunct Associate Professor of Agricultural Economics at the University of Western Australia's School of Agriculture and Environment. He completed a Bachelor of Economics degree with Honours at Murdoch University in 1995 and from 1996 to 2004 worked as a Regional Economist for the Department of Agriculture, Western Australia. During this period, he completed a PhD with The University of Western Australia's then School of Agricultural and Resource Economics (1999–2001) and a postdoc at the Wye campus of Imperial College London (2003–2004). From 2005 to 2011, he worked as a Research Economist with CSIRO Entomology (later Ecosystem Sciences), and was a Visiting Fellow at The Australian National University School of Resources, Environment and Society. In 2012, he returned to Western Australia to take up a Senior Economist position with DPIRD. He hails from the south west corner of Western Australia, home to a diverse mix of agriculture and one of the world's biodiversity hotspots. His career has been spent studying what happens when species of plants and animals are introduced to new agricultural and ecological systems via trade and other pathways. Since 2002, he has authored over 40 peer-reviewed papers in a diverse range of ecology, economics and policy journals and well over 100 cost–benefit analyses and impact assessments for different tiers of government both within Australia and overseas. He was the President of the Western Australian Branch of the Australian Agricultural and Resource Economics Society in the past and recipient of the CSIRO Medal for Research Achievement.

Janet Haddock-Fraser is a Professor of Sustainability and Leadership. She completed her first degree in Geography at the University of Oxford, before gaining commercial experience in the UK and Hong Kong. While in Hong Kong, she completed a master's degree and a PhD in Environmental Management at the University of Hong Kong, with the focus of her doctorate being the WTO and environmental trade disputes. Since this, she has worked as an academic at Imperial

College, London and the University of Kent, moving into senior leadership positions and the Professoriate at Canterbury Christ Church University and Manchester Metropolitan University. Her research profile is a diverse one with the common theme being the assessment of, and motivations for, sustainable behaviour.

Contents

Chapter 1

Introduction

1.1. Why Read This Book?

We live in a world that is increasingly dependent on international trade in light of substantial regional/national political tensions. Adding to this is an emerging understanding and concern about the social impact of biosecurity and ecosystem services risks associated with such trade. As the key international trade 'arbiter', the World Trade Organization (WTO) has never before faced such complexity within its decision-making remit.

This book provides an original analysis of these linked developments. In addition, it delivers a timely contribution to resolving environment-related international trade disputes. Specifically, this contribution provides a clear roadmap for improving WTO trade dispute resolution procedures so that both biosecurity and ecosystem services risks are taken into account in evaluating the social, economic and environmental impacts of international trade proposals and in so doing deliver enhanced multilateral social welfare.

This is important not just to enable effective decision-making for environment-related international trade disputes but also for the WTO as an organisation. With increasing numbers of bilateral and regional agreements, as well as new developments emerging such as the Comprehensive and Progressive Agreement for Trans-Pacific Partnership (CPTPP) initiated by multinational corporations in 2018, the WTO needs to implement ways of reinforcing its legitimacy and enhancing its relevance.

1.2. Background to This Book

The General Agreement on Tariffs and Trade (GATT) was established in 1948 with a legal agreement signed by 23 countries. Its primary objective was to promote international trade by reducing or eliminating trade barriers between countries. The GATT was succeeded by the WTO in 1995 with a legal agreement signed by 123 countries and with only minor modifications to the original GATT Agreement to promote international trade.

It is generally acknowledged that the overall effect of the GATT and the WTO since 1948 has been to reduce tariffs on imported goods and thereby encourage trade between countries (Goldstein *et al.*, 2007). However, such changes in international trading conditions were not expected to be straightforward or harmonious.

As a consequence, Article XX of the GATT Agreement was included to allow for 'General Exceptions', whereby a country could legitimately 'protect' itself from imports from other countries. In addition, a Dispute Settlement Understanding (DSU) was included to facilitate the settlement of disagreements between countries arising in the context of Article XX. With the subsequent establishment of the WTO came the 'Agreement on the Application of Sanitary and Phytosanitary Measures' typically referred to as the SPS Agreement. This agreement was included as an additional justification for preventing imports in order to protect human, animal or plant life or health from risks associated with international trade. As a consequence, the dispute settlement process could also be used to resolve trade disputes arising in the context of the SPS Agreement.

1.3. Motivation for and Objective of This Book

Our motivation for writing this book has been based on our perception of shortcomings in the GATT/WTO's existing procedures for evaluating and resolving international trade disputes which feature increased biosecurity and ecosystem services risks. Our view is that these procedures do not deliver on their objective of enhancing multilateral social welfare. Moreover, we think that these procedures can be improved so that the consequent decisions in the context of environment-related trade disputes can consistently enhance multilateral social welfare.

Given this motivation, our broad objective in this book is to demonstrate how the WTO's international trade dispute resolution procedures in the context of trade-related biosecurity and ecosystem services risks can be improved so as to enhance global social, economic and environmental welfare.

More specifically, we aim both to identify the failings of the WTO's existing procedures for evaluating international trade disputes which feature increased biosecurity and ecosystem services risks and to present a new methodological approach for evaluating the social, economic and environmental benefits and costs of international trade proposals which feature such environment-related risks. In so doing, we will argue the case for modifying the WTO's existing approach to international trade dispute resolution in the context of biosecurity and ecosystem services risks so that their decisions can consistently enhance multilateral social welfare.

1.4. Outline of the Contents of This Book

This book is structured into four parts. In broad terms, Part A, which comprises Chapters 2–4, contains a history and analysis of the GATT/WTO in the context of the resolution of environment-related international trade disputes. This part follows the format of Haddock (2000) in providing details of the WTO's existing environment-related trade dispute procedures, as well as undertaking an analytical critique of applications of these procedures to recent trade disputes, according to the methodology outlined in Haddock (2000).

More specifically, Chapter 2 contains a brief history of the GATT/WTO, including Article XX and the SPS Agreement, and its trade dispute resolution procedures. Chapter 3 contains details and critical analysis of global trade dispute applications made in the context of Article XX and the SPS Agreement that have led to the use of the GATT/WTO trade dispute resolution procedures in the context of biosecurity risks. Finally, in Part A, Chapter 4 contains details and critical analysis of global trade dispute applications made in the context of Article XX and the SPS Agreement that have led to the use of the GATT/WTO trade dispute resolution procedures in the context of ecosystem services risks.[1]

Part B, which comprises Chapters 5 and 6, builds on the critique of existing WTO trade dispute procedures relating to biosecurity risks presented in Chapter 3. These two chapters are focused on developing and

[1]In its broadest sense, ecosystem services include provisioning services (as well as regulating and cultural services) — such as food production — which deliver economic benefits to human welfare. However, in what follows, we consider provisioning services risks to be dealt with by the market-based analysis of Part B. As a consequence, Part C is focussed largely on risks to regulating and cultural services, which collectively deliver environmental and social benefits to human welfare (such as biodiversity and human health). This distinction also provides the basis for the separation of the analysis in Chapters 3 and 4.

applying a methodology for evaluating trade disputes which takes account of producer, consumer and biosecurity impacts in order to deliver a dispute resolution which enhances multilateral social welfare.

In particular, Chapter 5 outlines the methodology developed in Cook and Fraser (2008), which both identifies and enables the evaluation of the key impacts of the disputed trade so that the best decision for multilateral social welfare can be made in relation to that trade. Chapter 6 then presents two case studies of the application of the methodology outlined in Chapter 5. These two case studies draw from those trade disputes identified in Chapter 3 and illustrate the benefits of the proposed change in methodology for evaluating trade disputes involving biosecurity risks.

Part C, which comprises Chapters 7 and 8, builds on the critique of existing WTO trade dispute procedures relating to ecosystem services risks presented in Chapter 4. Similar to the two chapters of Part B, these two chapters are focused on developing and applying a methodology for evaluating trade disputes which takes account of producer, consumer and ecosystem services impacts in order to deliver a dispute resolution which enhances multilateral social welfare.

In particular, Chapter 7 outlines the methodology developed in Cook and Fraser (2015), which identifies and then considers evaluating the key impacts of the disputed trade with a view to making the best decision for multilateral social welfare in relation to that trade. In addition, Chapter 8 then presents two case studies of the application of the methodology outlined in Chapter 7. These two case studies draw from those trade disputes identified in Chapter 4 and illustrate the benefits of the proposed change in methodology for evaluating trade disputes involving ecosystem services risks. Note that the central difference between Parts B and C is that in the case of Part B, the impacts are largely economic and therefore lend themselves to standard economic evaluation. Whereas in Part C, the impacts are largely social and environmental and therefore typically non-monetised, which creates a different evaluation problem to be solved.

Part D of this book is entitled "A Way Forward", and here our vision is presented in two chapters. Chapter 9 reflects on the complexities which have been revealed in the preceding analysis — particularly those relating to dealing with uncertainty, with non-monetised impacts, and with distributional considerations. In so doing, we suggest ways of dealing with such complexities in order to facilitate good decision-making, including the use of a consensus-building methodology such as multi-criteria decision analysis (MCDA).

Finally, Chapter 10 provides an integration of the methodological modifications proposed in this book. These proposals collectively represent our envisaged contribution towards improving WTO dispute resolution procedures. Specifically, we propose that both biosecurity and ecosystem services risks are taken into account in evaluating the social, economic and environmental impacts of disputed international trade, so as to enable the WTO's resolution procedures to deliver decisions that consistently enhance multilateral social welfare.

Part A
History and Analysis of the WTO in the Context of Trade Disputes

Chapter 2

Brief History of the WTO and Its Environment-Related Trade Dispute Resolution Procedures

2.1. Introduction

The World Trade Organization (WTO) represents the global voice for maximisation of economic welfare through 'free trade' or 'least restrictive trade'. It was launched in 1995, emerging from the 1994 Uruguay Round of its predecessor, the General Agreement on Tariffs and Trade (GATT), with a superset of responsibilities and authority. The GATT was felt to be deficient in its scope, in particular in the following:

- It included primary and manufactured goods but not services or intellectual property.
- Any GATT decision was advisory, whereas the WTO brought authority and mandate to compliance with its measures.
- The GATT objectives were felt to be limited in their incorporation of the modern understanding of economic welfare.

With regard to this final issue, although both the GATT and WTO call for the maximisation of economic welfare, the latter explicitly recognises wider consideration of environmental and social welfare within the scope of economic welfare. The objectives state as follows:

> *[Members'] relations in the field of trade and economic endeavour should be conducted with a view to raising standards of living, ensuring full employment and a large and steadily growing volume*

> *of real income and effective demand, and expanding the production of goods and services, while allowing for the optimal use of the world's resources in accordance with the objectives of sustainable development, seeking both to protect and preserve the environment and to enhance the means or doing so in a manner consistent with [the members'] respective needs and concerns at different levels of economic development.*
>
> (WTO preamble, 1995, p. 1)

In this chapter, we provide a brief history of the WTO in general before exploring its mandate, regulations and decisions relative to the widely defined issue of 'environment'. The first section considers the history of international trade regulation in relation of the development of the GATT and WTO including the objectives, cultural and socio-political context of each. The second section explains the decision-making mechanism used, through a quasi-legal, rules-based approach. We follow this with a section which explains briefly the WTO Dispute Settlement process, used as the key means used to resolve trade–environment disputes. The fourth section reviews how the subject of 'environment' has been included within the WTO (as well as its predecessors and relevant international agreements). In particular, it will consider the role of GATT Article XX (subsumed into WTO regulations) and the Sanitary and Phytosanitary Standard (SPS). The former has been the principal means through which ecosystem services decisions on trade disputes have been considered, and the latter has been used for biosecurity cases. In the final section, we critique the method currently used to resolve trade—environment disputes, highlighting the gaps in the process used, which result in sub-optimal decision-making.

2.2. Historical Context of the World Trade Organization

In order to understand why the predecessor to the WTO, the GATT, was established, we need to understand the economic conditions and circumstances that existed prior to its development. These started after the First World War (1914–1918) in Europe when many nations sought to gain economic advantage by the use of protectionist measures as a defensive measure. This was in part caused by their economies having become disorganised, their productive facilities disrupted and their channels of trade broken as a result of the war. Many nations had incurred heavy debts resulting in feelings of increased nationalism and protectionism (Wilcox, 1949, pp. 5–9).

The impact of such behaviour reached a critical point at the time of economic recession in the late 1920s, when there was a sharp contraction in the volume of world trade. Because of this, the US enacted the Hawley–Smoot tariff in 1930, raising duties to 52%, the highest in US history (Esty, 1994, p. 243). Such behaviour is considered by historians to have tipped the economic downturn of the late 1920s into the global Great Depression of the 1930s. Conjecture suggests this was, in part, responsible for the subsequent conditions that led to war in Europe and the Pacific at the end of the 1930s (Esty, 1994, p. 244).

The necessity for economic and trade cohesion was identified during this period though. The League of Nations organised various conferences in the 1920s and 1930s, stressing in principle the need for free trade (e.g. at Geneva in 1927 and Lausanne in 1932) (League of Nations Secretariat, 1945, pp. 118–124). These efforts, though worthy, did not deliver a practical means forward.

Coming out of the Second World War, many trading nations felt that it would be an error to repeat the economic and trade circumstances of the 1920s and 1930s. Hudec suggested that the postwar design for an international trade policy and regulating institution was brought about by single-minded concern among the major trading nations to avoid repeating such errors (Hudec, 1990, p. 5). Additionally, the implications of the Bretton Woods Agreement which saw the establishment of the International Monetary Fund among other institutions, made more countries amenable to exposing their economies to foreign competition (Shutt, 1985, pp. 12–13).

Discussions on the need for a free trade mechanism began as early as 1943, which formed the basis for a proposal for an International Trade Organization (ITO). There were plans for rigorous codes of conduct which included absolute prohibition on all forms of non-tariff trade restrictions (except for when quantitative restrictions were put in place in times of balance of payments crisis), as well as the elimination of export subsidies and the progressive reduction in tariffs (Hudec, 1990, p. 15). The ITO in this complete form never came to fruition, with concerns from many about mandated free trade obligations. The lawmakers in the US, in particular, seemed reluctant to cede sovereignty to an international organisation, despite the fact that they had to that point been at the forefront of proposals for a free trade mechanism (Esty, 1994, p. 244).

Although the ITO proposal required ratification of each state to become a legally binding instrument, the supporting document (which was never achieved), the GATT did not require ratification in order to be adopted. It was deemed to be an 'executive agreement' not requiring legislative approval at national government level (e.g. by the US Congress). As such, the GATT came into 'force' in January 1948 and was expected to provide a temporary framework for trade liberalisation efforts pending the passage of the ITO (Esty, 1994, p. 245). Legally, the GATT was a trade agreement, which members (governments) agreed to accept provisionally. Despite this, many considered to have some legal authority and as such is often considered as a treaty in all but name.

The GATT system for managing the free trade mechanism remained substantially the same until 1995. Major change occurred as a result of the Uruguay Round of GATT in 1994 once there was finally an agreement among members to set up a formal trade organisation to act as an umbrella administration system for GATT and other elements of the international trading system. The organisation formed was the World Trade Organization (WTO). GATT was subsumed into the WTO, but the structure and many of the rules of the GATT continue to exist substantially as they were prior to the formation of the WTO.

The rationale for the move to the WTO was driven by a multitude of factors. It had become apparent by the 1980s that the GATT was no longer relevant to the realities of world trade as it had been in the 1940s. World trade had become more complex and important. The globalisation of the world economy was underway and international investment was increasing very rapidly. The trade in services was becoming more important but was not covered by GATT. There were also concerns about GATT's ability to deal with liberalising trade in agriculture, textiles and clothing, as well as questions regarding the effectiveness of the institutional structure and dispute resolution mechanisms.

The scope of the GATT was originally designed to cover free trade in mercantile goods, such as agricultural, mined, fished and manufactured goods. The inception of the WTO saw the additional inclusion of rules governing services and intellectual property protection. Such areas were of little concern when the GATT was initially set up but are now a significant part of trade.

Neither the GATT's nor the WTO's objectives are free trade *per se*, but economic growth, with free trade (or least restrictive trade) being the mechanism by which to achieve this. The rules of the GATT detail how trade restrictiveness should be reduced and assume that free trade is the mechanism by which economic growth occurs. As such, the rules are not rules on 'how mutual economic growth can be maximised', but merely 'how trade restrictiveness should be reduced'.

2.3. Rules-Based Decision-Making

The objectives of GATT and WTO are achieved through adherence to a number of rules or articles. The principles to which the rules refer call for the following between members (being nation states):

- non-discrimination;
- reciprocity;
- binding and enforceable commitments (since WTO inception);
- transparency;
- 'safety' provision within 'least-restrictive' trade to protect, for example, the environment, human, animal or plant health.

The substance of many articles has remained largely unchanged since their initial drafting in the 1940s although there have been some amendments and additions to them over time. There are many texts providing substantial detail and discourse on the numerous articles that make up the jurisdictional framework for the WTO (e.g. Baughen, 2007). In this book, we will focus on only those most relevant to dispute resolution decision-making for ecosystem services and biosecurity trade disputes. Notwithstanding this, it is appropriate to highlight two cornerstone articles which have endured within the system: Article I (most favoured nation principle) and Article III (national treatment principle). The former obliges a member to extend to any other member any privilege of advantage which it extends to like products which are destined for or imported from other countries. The latter provides that the products of the territory of any contracting party imported into the territory shall be accorded treatment no less favourable than that accorded to like products originating in the importing territory. Other articles set the terms for use of tariffs and the means by which disputes should be dealt with. Of particular interest here are Article XX (General Exceptions) and the Sanitary and Phytosanitary Standards (SPS) because of their relevance to trade–environment disputes. These will be covered later in Section 2.5.

As noted above, the WTO is built on adherence to its 'rules' under the aegis of international treaty law. There is much discussion in the legal community about the precise nature of the WTO (and GATT) agreement (and the nature of international law generally), but for the purposes of this book, we have limited our discussion here to the relevance of this 'rules-based' approach to decision-making in more general terms.

Decision-making through the 'rules-based' approach has two components: the law itself and 'legal' reasoning. The law itself is a collection of articles, rules, principles, policies and (if appropriate to the jurisdiction) precedents employed by the decision maker to justify their decision (Burton, 1995). Legal reasoning provides the process used to present legal argument. It can be presented as a step-based approach (with some circularity) as follows (Vandevelde, 1996):

- identify applicable sources of law, usually statutes or judicial decisions;
- analyse these sources of law to determine the applicable rules of law and the policies underlying those rules;
- synthesise the applicable rules of law into a coherent structure;
- research the available facts;
- apply the structure of the rules to the facts to ascertain the rights or duties created by the facts, using the policies underlying the rules to resolve difficult cases.

Legal reasoning within the WTO is complicated by the nature of the law itself (outside the bounds of domestic jurisdiction and operating internationally, as a form of international treaty), and its boundaries and authorities. It is not uncommon for WTO members to dispute whether certain behaviour does or does not conform to rules within, or beyond, its jurisdiction.

In general, the content of WTO disputes is centred on the legal interpretation of its rules. Three fundamental approaches have been identified as possible means for interpretation in this manner (Yambrusic, 1987). These are 'clear sense', 'interpretation' and 'context':

- Clear sense suggests the rules as they are written are how they should be read from a legal viewpoint. While conceptually neat, in practice the likelihood of the intended meaning or phrase meaning exactly the same in the many differing languages or cultures represented by WTO membership is fanciful.
- Interpretation lies at the other extreme to 'clear sense', recognising that the purpose of a rule may change over time as experience is gained in its operation and use. This approach parallels the notion of 'precedent'

to the extent that the approach to applying the rule may modify as a result of experience, but is rife with inconsistency, not least as the use of precedential approaches to law differ across WTO member jurisdictions.

- The contextual approach seeks to arrive at the meaning of words in a 'rule' by considering all the relevant factors of the particular situation and applying it on that basis. This approach calls for much more information/data of relevance to a particular dispute to be provided to aid decision-making and supports the approach called for in this book. An issue here is how to establish what data are relevant in a trade–environment dispute, where matters of trade can impact on social, environmental and economic factors.

2.4. Dispute Resolution

The dispute settlement process in the WTO is used where one or more parties believe fellow WTO members are violating trade rules in a manner that affects the complaining party directly and wish to seek resolution of this. A procedure for settling disputes was in place under GATT and the means to do this through reference to Article XXIII and Article XXII of the GATT as a starting point for dispute settlement is still relevant today. Article XXIII is the 'nullification and impairment article'. This basically states that if any contracting party considers the benefits accruing to it under GATT/WTO are being nullified or impaired by the behaviour of any other contracting party, it can bring the complaint forward for resolution. Article XXII provides the mechanisms and means for reasonable consultation during the process of dispute settlement, with all parties directly affected and potentially third parties.

While a procedure for settling disputes existed under the GATT, it did not carry the legislative might of the WTO. GATT rulings could only be adopted by consensus, meaning that a single objection could block the ruling and the processes of dispute resolution developed through custom and precedent. There were no fixed timetables for resolution and with the 'advisory' nature of the rulings, they were easier to block or delay indeterminately (WTO, 2018a). With the formation of the WTO, a more structured process with clearly defined stages and deadlines was introduced, although the substance of how disputes are resolved (i.e. through a Panel process scrutinising the case with reference to WTO articles and rules) remains intact. As with GATT, resolution outside of the dispute settlement process, through consultation and mediation, is sought and preferred where possible. This is, not least, because even with the 'speedier' WTO system, a formal dispute may take 12–15 months from setup to settlement (WTO, 2018d).

Figure 2.1 shows a detailed flowchart of the WTO dispute settlement process, making reference to the implementation of relevant articles of the Understanding on the Rules and Procedures Governing the Settlement of Disputes and timescales by which each element of the process should normally be completed.

Of particular interest in the trade–environment dispute resolution process (if context is a consideration in decision-making) is the opportunity to engage an expert review group (through Article 13). Article 13 provides the 'Right to Seek Information' and technical advice as appropriate from either individuals or other organisations and may be in the form of a written advisory report from an expert group. Information cannot be sought from an expert review without the knowledge of the member under consideration in the dispute, but any confidential information provided by and expert

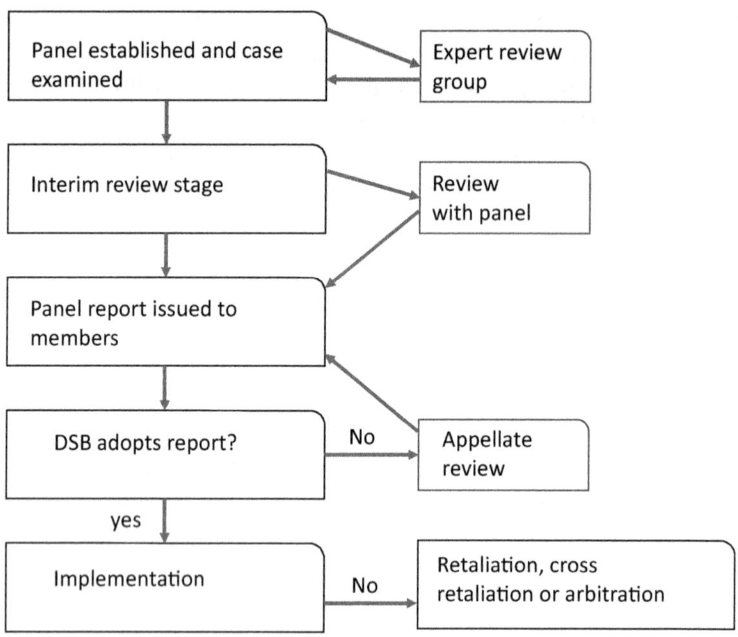

Figure 2.1. WTO dispute settlement process.
Source: Adapted from: https://www.wto.org/english/thewto_e/whatis_e/tif_e/disp2_e.htm accessed 9th January 2018.

review group cannot be revealed without the formal authorisation from the individual, body, or authorities of the member providing the information. Appendix 4 in the dispute resolution process provides procedures and processes for the establishment and operation of the expert review groups. Since the inception of the WTO dispute resolution process, Article 13 has been triggered in both ecosystem services and biosecurity trade disputes including the following:

- US–Netherlands and EU — Tuna-Dolphin II, 1994, DS29/2, DS29/3, ILM 839.
- DS18 — Australia vs. Canada — Measures affecting importation of salmon – 1995.
- US vs. India, Pakistan, Malaysia and Thailand — Shrimp and Shrimp products, 1998, DS58.
- EU vs. Brazil — Retreaded Tyres, 2007, DS332.

2.5. Rules Pertaining to Trade and Environment

Despite its original remit to encourage free trade principles, the WTO has made some exceptions to it. These include the 'Special and Differential Treatment' (S&D) provisions for less developed countries. They provide measures to increase trading opportunities for developing countries and for all WTO members to safeguard the trade interests of developing countries. For the most part, exceptions within the WTO are articulated and arbitrated through the General Exceptions Article (Article XX) and, of particular relevance to trade and environment, the SPS. Article XX has been used since the early days of the GATT but has been added to and incorporated into WTO rules. The SPS which consider trade exemptions for food-related health and safety reasons was introduced as part of the formulation of the WTO in 1995.

2.5.1. *Article XX*

Article XX has two parts: (i) a list of the types of exceptions for which members may legitimately take action counter to the other general principles of the WTO (items a–j) and (ii) the *chapeau* to the article that contextualises the use of these exceptions and highlights the conditions under which they may be used. The full wording of Article XX (WTO, 2018b) is shown in Figure 2.2, with the items relevant to trade–environment dispute settlement provided in italics for emphasis.

Subject to the requirement that such measures are not applied in a manner which would constitute a means of arbitrary or unjustifiable discrimination between countries where the same conditions prevail, or a disguised restriction on international trade, nothing in this Agreement shall be construed to prevent the adoption or enforcement by any contracting party of measures:

(a) necessary to protect public morals;

(b) *necessary to protect human, animal or plant life or health;*

(c) relating to the importations or exportations of gold or silver;

(d) necessary to secure compliance with laws or regulations which are not inconsistent with the provisions of this Agreement, including those relating to customs enforcement, the enforcement of monopolies operated under paragraph 4 of Article II and Article XVII, the protection of patents, trade marks and copyrights, and the prevention of deceptive practices;

(e) relating to the products of prison labour;

(f) imposed for the protection of national treasures of artistic, historic or archaeological value;

(g) *relating to the conservation of exhaustible natural resources if such measures are made effective in conjunction with restrictions on domestic production or consumption;*

(h) undertaken in pursuance of obligations under any intergovernmental commodity agreement which conforms to criteria submitted to the CONTRACTING PARTIES and not disapproved by them or which is itself so submitted and not so disapproved;*

(i) involving restrictions on exports of domestic materials necessary to ensure essential quantities of such materials to a domestic processing industry during periods when the domestic price of such materials is held below the world price as part of a governmental stabilization plan; Provided that such restrictions shall not operate to increase the exports of or the protection afforded to such domestic industry, and shall not depart from the provisions of this Agreement relating to non-discrimination;

(j) essential to the acquisition or distribution of products in general or local short supply; Provided that any such measures shall be consistent with the principle that all contracting parties are entitled to an equitable share of the international supply of such products, and that any such measures, which are inconsistent with the other provisions of the Agreement shall be discontinued as soon as the conditions giving rise to them have ceased to exist.

*Interpretative Note Ad Article XX from Annex I, Sub-paragraph (h). The exception provided for in this sub-paragraph extends to any commodity agreement which conforms to the principles approved by the Economic and Social Council in its resolution 30 (IV) of 28 March 1947.

Figure 2.2. Article XX: General exceptions.

Source: WTO (2018b).

Reviewing these from the disciplinary perspective of economics, a number of the exceptions provided relate to the amelioration of subsidies (or their inverse, externalities) in either a societal and/or environmental context. Articles XX(a), (e) and (f) speak to these societally either through discouragement of prison (i.e. free) labour (although interestingly not slave or child labour) or through recognition of the social worth of ethical behaviour and preservation of cultural artefacts. With regard to trade–environment disputes, the *chapeau* and items (b) ('necessary to protect human, animal or plant life or health') and (g) ('relating to the conservation of exhaustible natural resources if such measures are made effective in conjunction with restrictions on domestic production or consumption') are those that have been used, particularly for ecosystem services disputes, including the iconic Tuna–Dolphin cases between the US and Mexico.

A total of seven trade–environment cases were considered under Article XX prior to 1995 (so under GATT rather than WTO) and four since (between 1995 and 2018). It is generally considered that the GATT Panels were somewhat restrictive in their interpretation of the rules, particularly in terms of (i) applying strictures to the term 'necessary' and how burden of proof could be provided, and also from (ii) taking the stance that environmental-related trade restrictions could only be legitimised if they had a direct impact only within the territory of the complainant. Thirdly, process and production methods which have a deleterious environmental impact were not to be included (Baughen, 2007). Since the formulation of the WTO, these strictures have been loosened with Panels now seeking to balance trade and environmental impacts (Saito, 2001) rather than protection of least trade restrictiveness as a priority. Saito conjectures this has arisen by adherence to the WTO preamble, seeking greater consideration of sustainable development and resource preservation, compared to the more polemic GATT preamble. Notwithstanding this, decision-making in dispute settlement is still dominated by a jurisprudential approach.

2.5.2. *Sanitary and Phytosanitary Standards (SPS)*

The SPS agreement was introduced as part of the launch of the WTO in 1995. It covers the measures applied to protect against a range of specified risks to animal, plant or human health through trade in food. Thus, it is concerned with food safety and animal/plant health and welfare. In the 1947 GATT, exemptions to free trade existed (through Article XI), enabling quotas on agricultural products, which led to market distortions. The SPS seeks to rectify these, while protecting importing countries against health and safety issues that may arise through the introduction of exotic disease, invasive species or additives/pesticides not permitted in the importing country.

The SPS shares many of the core principles that apply to Article XX exemptions, in particular the need to ensure restrictions on trade are neither arbitrary nor discriminatory and that the measure is the most appropriate while ensuring least trade restrictiveness. Additionally, it specifies the need for the protection (to animal, human, plant life and health) considered should only be within the territory of the member concerned. The agreement does allow for differential standards to be applied at a domestic level, as long as that standard is applied in a non-discriminatory manner with other members.

The introduction of the SPS was considered to be a measure which benefited (or did not adversely affect) consumers, developing countries, exporters or importer (WTO, 2018c). It is claimed that consumers benefit by the assurance (and possibly enhancement) of food safety through not only the systematic use of scientific information in trade decisions but also the elimination of unnecessary trade barriers, so allowing consumers to benefit from a 'greater choice of safe foods and from healthy international competition among members'. Developing countries benefit from the availability of the international framework for SPS and the availability of increased technical assistance. Exporters and importers are both thought to benefit from the elimination of unjustified barriers and greater certainty about the conditions expected from other markets.

Unlike Article XX, the SPS introduces two new notions: evaluation of risk and use of scientific evidence. With regard to the former, this is articulated through Article 5 of the SPS and the WTO introduction to the SPS notes as follows:

> *An acceptable level of risk can often be achieved in alternative ways. Among the alternatives — and on the assumption that they are technically and economically feasible and provide the same level of food safety or animal and plant health — governments should select those which are not more trade restrictive than required to meet their health objective. Furthermore, if another country can show that the measures it applies provide the same level of health protection, these should be accepted as equivalent. This helps ensure that protection is maintained while providing the greatest quantity*

> *and variety of safe foodstuffs for consumers, the best availability of safe inputs for producers, and healthy economic competition.*
>
> (WTO, 2018d)

Article 5 requires members to base measures on an assessment 'as appropriate to the circumstances of the risks to human, animal or plant life or health, taking into account risk assessment techniques developed by the relevant international organisation'. Of particular interest in the context of this book are the economic-trade implications of the SPS. In paragraph 3 (Article 5), it states that in assessing risk members should take account of the following (Cook and Fraser, 2008):

- the potential damage in loss of production or sales in the event of the entry;
- the establishment or spread of the pest or disease;
- the cost of control or eradication in the territory of the importing member;
- the relative cost effectiveness of alternative approaches to limiting risk.

Article 5 also specifies that any restrictions placed on imported goods to reduce expected producer losses should be justifiable by way of a risk analysis and that the methodology used for this analysis draw upon internationally recognised methodologies.

With regard to scientific evidence, this is covered by Article 2 of the SPS. This requires that measures must be 'based on scientific principles' and 'not maintained without sufficient scientific evidence' stating as follows:

> *Members shall ensure that any sanitary or phytosanitary measure is applied only to the extent necessary to protect human, animal or plant life or health, is based on scientific principles and is not maintained without sufficient scientific evidence, except as provided for in paragraph 7 of Article 5.*
>
> (WTO, 2018d)

The rider mentioned (paragraph 7 of Article 5) provides for members to provisionally adopt sanitary or phytosanitary measures on the basis of information available if relevant scientific evidence is not available, including use of information from relevant international organisations.

The SPS, through its call for risk assessment and use of scientific evidence to be used within any decision framework, suggests the importance of the context-led approach to applying the SPS rules. Notwithstanding the wider requirement of the WTO to ensure non-discrimination, least-trade restrictiveness, and the geographical restrictions on responsibilities (territoriality), the SPS represents a step away from decision-making through the interpretation of a set of rules as the only means to arbitrate on trade disputes.

2.6. Alternative Models for Decision-Making

As supporters of a context-based (i.e. data-based) approach to decision-making, the opportunity within the SPS to look at risk and scientific evidence is heartening, as is the WTO's desire to 'balance trade and environmental factors' when arbitrating disputes using Article XX and the inclusion in the Dispute Settlement process enabling the use of expert review groups. However, these measures of themselves are not sufficient to provide an understanding of the inherent complexity of issues surrounding a trade–environment dispute, nor to establish whether sustainable development (as called for in the WTO preamble) is optimised. Such partial analysis risks a decision being made that results in sub-optimal social, political, economic or environmental consequences on the members directly impacted, or even third parties (Haddock, 2000).

Further to this concern, Haddock (2000) proposed the use of a data-driven approach to decision-making for trade–environment disputes, noting that such an approach would require the consideration of four major system components as follows:

1. The impact of trade on the natural environment, as well as the impact of any environmental change with reference to its impact on the economy, technology and society should be measured.
2. Where trade–environment disputes involve the use of technology (e.g. gasoline cases, retreaded tyres, use of genetic modification), an understanding of the science *per se* is needed (possible now for biosecurity cases in SPS but not automatically for trade–environment cases under Article XX). Additionally, the impact of technology on society, environment and economy should be considered.
3. The optimal outcome of any trade–environment dispute should reflect the values and goals of society at large, both at nation-state level and for the individuals that make up society. Value and cost should be measured to assess impact where possible.

As such, the impact on trade (least trade restrictiveness) should be measured against economic impact (producer/consumer/impact on community), environmental economic impact (value/loss of environmental amenity to humans), and impact on the natural environment in itself. Only in balancing these attributes and having sufficient data on each to make an informed judgement can the WTO meet its objective of maximisation of social and economic welfare while meeting the objectives of sustainable development.

Cook and Fraser (2008) adopt a similar stance on the need for a data-driven approach for biosecurity cases. They model decisions on the import of apples into Australia from New Zealand based, first on the WTO dispute resolution approach (using SPS), and second, using unilateral welfare-maximising approach which considers the consumer gains from trade, the loss of domestic producers' market share and expected damage from the invasive species. The latter models impact of a decision whereas the former seeks adherence to the rules-based approach and demonstrates the possibility of a different outcome under each system.

Both approaches call for a holistic scope of impact to be considered in decision-making, and a data-based, evidence-based and value-driven approach to evaluating the impact of a trade–environment decision, considering producer, consumer, environmental impact and wider economic (societal) impact in the round. The use of Article XX, and/or SPS, are not able to deliver this in their current articulation, nor are the current WTO dispute resolution processes equipped to operate on this basis.

The following two chapters will assess, first, biosecurity disputes (Chapter 3) and, then, ecosystem services disputes (Chapter 4) against the wider evaluation framework suggested above to highlight the extent to which the dispute resolution Panels have (or have not) considered such impacts in their decision-making.

Chapter 3

WTO Trade Dispute Resolution Procedures in the Context of Biosecurity Risks — Details and Critical Analysis of Applications to Global Trade Disputes

3.1. Introduction

In this and the following chapter, we describe and evaluate specific GATT/WTO disputes relating to biosecurity (this chapter) and ecosystem services (Chapter 4). The purpose of this is to demonstrate not just how decision-making has been made with respect to biosecurity and ecosystem services disputes but also to highlight the limitations of the approach adopted, particularly in relation to the lack of engagement with data relevant to understanding the wider social, environmental and economic impacts of the case. In Chapter 2, we argued this was needed to enable the WTO to meet its core mission of least trade restrictiveness while delivering on sustainable development and social welfare.[1]

In this chapter, we concentrate on biosecurity. This has been defined as follows:

> *a strategic and integrated approach to analysing and managing relevant risks to human, animal and plant life and health and associated*

[1]Social welfare is used to encompass human wellbeing not just of economic impacts but also environmental and social as consistent with Article XX and definitions of sustainable development.

risks for the environment.... the overarching goal is to prevent, control and/or manage risks to life and health as appropriate to the particular biosecurity sector. In doing so, biosecurity is an essential element of sustainable agricultural development.

(World Health Organization, 2010a)

Significantly, biosecurity matters relate to human food production and thus could more readily be quantified with respect to cost and value than the ecosystem services cases to be considered in Chapter 4. As shown in Table 3.1, biosecurity hazards come in a variety of forms, affecting both plants and animals.

We discuss below on a case-by-case basis how the WTO has handled each biosecurity dispute to date. We have identified these are being those cases where some element of hazard (in Table 3.1) as applied to agricultural products has been identified as a subject in the trade dispute. The cases are all post 1995 and the introduction of the WTO Agreement on the

Table 3.1. Definitions of hazard as applicable to different biosecurity sectors.

Sectors	Definitions of Hazard
Food safety	A biological, chemical or physical agent in, or condition of, food with the potential to cause an adverse health effect.
Zoonoses	A biological agent that can be transmitted naturally between wild or domestic animals and humans.
Animal health	Any pathogenic agent that could produce adverse consequences on the importation of a commodity.
Plant health	Any species, strain or biotype of plant, animal or pathogenic agent injurious to plants or plant products.
Plant health quarantine	A pest of potential economic importance to the area endangered thereby and not yet present there, or present but not widely distributed and being officially controlled.
'Biosafety' in relation to plants and animals	A living modified organism that possesses a novel combination of genetic material obtained through the use of modern biotechnology that is likely to have adverse effects on the conservation and sustainable use of biological diversity, taking also into account risks to human health.
'Biosafety' in relation to food	A recombinant DNA organism directly effecting or remaining in a food that could have an adverse effect on human health.
Invasive alien species	An invasive alien species outside its natural past or present distribution whose introduction and/or spread threatens biodiversity.

Source: World Health Organization (2010a).

Application of Sanitary and Phytosanitary (SPS) Measures. This was introduced as one of a range of new measures (and dispute settlement processes) upon the advent of the WTO.

Each case is structured so that the first section summarises the case brought to dispute settlement, providing a brief description in relation to the biosecurity component. The second section provides an analysis of the core arguments used by the dispute Panel relating to this with the third section exploring how 'data' or 'facts' relating to the impact of the situation under dispute have been used in the decision-making. This information is broadly categorised into scientific, trade and economic impacts, and the analysis highlights not only the extent of information considered but also whether it formed part of the WTO dispute Panel evaluation of the case. The final section provides the outcome from the dispute settlement Panel. The chapter concludes with a discussion of the processes and precedents used and the opportunities missed by omission of a data-centric evaluation component in the context of the WTO's stated aim of resolving disputes based on the objective of maximising social welfare.

3.2. Biosecurity Disputes

There has been a total of six disputes which have been evaluated through the dispute settlement system in the realm of biosecurity since the incorporation of the WTO in 1995:

1. DS18: Australia vs. Canada — Measures affecting the importation of salmon (1995)[2];
2. DS245: Japan vs. USA — Measures affecting the importation of apples (2002);
3. DS367: Australia vs. New Zealand — Measures affecting the importation of apples from New Zealand (2007);
4. DS430: India vs. USA — Measures concerning the importation of certain agricultural products (2012);
5. DS447: Argentina vs. USA — Measures affecting the importation of animals, meat and other animal products from Argentina (2012);
6. DS475: Russian Federation vs. EU — Measures on the importation of live pigs, pork and other pig products from the European Union (2014).

[2]The date provided is that when the request for consultations were received by the WTO.

There have been many other biosecurity disputes brought to the WTO, but these have been excluded from this analysis as they either (i) did not go through a Panel dispute settlement process as a mutually agreed decision between the parties was adopted before the case was heard by the WTO, or (ii) the case is 'live' in that the Panel has yet to formulate a decision. In this latter category, there are currently (2019) 20 such cases 'in consultation', although some have been so for more than a decade. Each case here is a summary based on the Panel Report, Appellate Body Report (where relevant) and WTO dispute summary documents from the WTO website.

3.3. DS18: Australia vs. Canada — Measures Affecting The Importation of Salmon (1995)

3.3.1. *Introduction*

In 1975, Australia introduced, through the Australian Quarantine Proclamation, that salmon and salmonids should be prohibited from import unless they had undergone acceptable levels of treatment likely to prevent the introduction of any infection or contagious diseases. Heat-treated salmon products as well as non-commercial quantities of salmon (for scientific purposes) were accepted for import.

Following 20 years of import bans on Canadian salmon and associated products by Australia, Canada took forward their concerns through the WTO dispute settlement process. It contended that Australia's actions were in violation of GATT Articles XI (elimination of quantitative restrictions) and XIII (non-discrimination) of GATT (1994) and Articles 2 and 5 of the SPS Agreement and that the application of the Australian measures nullified or impaired benefits accruing to Canada pursuant to the WTO Agreement. These two SPS articles enable member countries' basic rights and obligations to take SPS measures necessary for the protection of human, animal or plant life or health, provided that such measures are not inconsistent with other GATT/WTO provisions (and include the use of scientific evidence and risk assessments outlined in Article 5 of SPS as well as other requirements of non-discrimination).

3.3.2. *Use of SPS and other relevant rules*

As part of the consultation and dispute resolution process and as required through Article 5 of SPS, Australia's agreed to document an Import Risk Analysis (IRA) on the quarantine issues involved with the import of uncooked salmon meat (in this instance wild, ocean-caught Pacific salmon from North America), with the 'final'[3] IRA provided in December 1996. This provided identification of potential disease agents, analysis of disease risks, scientific review of data for salmonid diseases, a socio-economic assessment of the potential impact of salmonid disease, and an introduction and identification of options for risk reduction and risk management.

Within this, it was noted that there was no evidence of the spread of diseases via fish products for human consumption, despite 'the wide scale movement of salmonid product within and between continents'. The draft report also explained the sequence of events required for imported salmon to cause an outbreak and establishment of any 'exotic' disease in Australia. It was noted that although zero risk was not attainable, the risk of disease introduction had a low probability. By the time the final report was produced, the conclusions had modified somewhat suggesting that there "*was a possibility that up to 20 disease agents exotic to Australia might be present in Pacific salmon products and although the probability of establishment would be low, there would be major economic impacts which could seriously threaten the viability of aqua-cultural operations and the recreational fishing industries, in addition to adverse environmental impacts on the built environment of Australia*". The report further noted that any disease outbreak from this would be impossible to eradicate. The report was accepted by the Australian Director of Quarantine who considered that imports of uncooked salmonid products from Canada and the US should not be permitted in light of the unique circumstances (assumingly with regard to Australian fauna and flora), risking the introduction of a range of potential disease agents and potential socio-economic and environmental impacts arising from these.

Canada disputed Australia's assertion that its import restrictions were acceptable within WTO rules. In particular, they considered the following:

- It was not justified under Articles XI:2 or XX of the GATT.
- While it falls within the scope of the SPS Article 1 definition, they considered it:
  - violated Articles 3.1 and 3.3 as it is not based on international standards, guidelines or recommendations;

[3]The word 'final' is used here to indicate the version considered by the Panel, but Australia subsequently released another IRA in July 1999 (Australian Quarantine and Inspection Service, 1999).

  - did not consider the risk assessment by Australia to be valid, asserting that Australia did not use the full range of scientific evidence available (Article 5);
  - considered that the measure was inconsistent with Australia's obligation to avoid arbitrary and unjustifiable distinctions and was a disguised trade restriction (Article 2.3).
- There were other less trade-restrictive means possible (Article 5).

Australia countered these assertions within its defence and additionally noted that the information provided to the Panel through the IRA included, as well as scientific data, economic and trade-related data in line with the requirements of SPS Articles 5.3 and 5.4:

- **SPS Article 5.3:** *In assessing the risk to animal or plant life or health and determining the measure to be applied for achieving the appropriate level of sanitary or phytosanitary protection from such risk, members shall take into account as relevant economic factors: the potential damage in terms of loss of production or sales in the event of the entry, establishment or spread of a pest or disease; the costs of control or eradication in the territory of the importing Member; and the relative cost-effectiveness of alternative approaches to limiting risks.*
- **SPS Article 5.4:** *Members should, when determining the appropriate level of sanitary or phytosanitary protection, take into account the objective of minimizing negative trade effects.*

3.3.3. *Use of data*

Substantial levels of data were provided to the Panel as part of this dispute resolution process. Most of it was scientific evidence, against which the two parties disagreed, with both employing experts to comment. The scientific data provided included the following:

- the number of disease agents at issue and the number exotic to Australia;
- the risk of entry and spread;
- specifics on each pathogen and the factors that affect its transmission.

Reference is also made to the need to understand the economic and environmental consequences of import, noting that it is important to consider both the economic asset value of salmonid aquaculture and recreational fisheries, and the biological value of native salmonids and their value in export markets. Australia noted that its IRA included a stochastic assessment of economic impact in relation to two of the diseases and the economic and technical feasibility of prevention and control. However, there is no mention within the Panel Reports of what this data was,[4] with the initial Panel noting that Canada did not challenge Australia's claim under SPS Article 5.3 relating to 'relevant economic factors' and so did not see reason to further examine this information.

3.3.4. *Outcome*

The original Panel conclusions for this case found that Australia's measures were inconsistent with Articles 2.2, 2.3, 5.1, 5.5 and 5.6 of the SPS Agreement and also nullified or impaired benefits accruing to Canada under the SPS Agreement and rejected the legitimacy of the IRA. Australia appealed this decision, and the Appellate Body concurred with the original Panel in that it considered Australia had acted inconsistently with the SPS.

3.4. DS245: Japan vs. USA — Measures Affecting the Importation of Apples (2002)

3.4.1. *Introduction*

In March 2002, the US brought the issue of import restrictions of apples from the US to Japan, with the restriction imposed by Japan allegedly as a means to protect the nation against fire blight. There were multiple detailed requirements relating to import restrictions from Japan which cumulatively the US felt violated the WTO. The requirements stipulated by Japan were as follows:

- Fruit must be produced in designated fire blight-free orchards through agreed procedures by the US Department of Agriculture. At the time of the dispute, only orchards in the states of Washington and Oregon were considered acceptable.
- The export orchard must be free of infected plans and free of host plants of fire blight (other than apples) regardless of whether they are infected or not.
- The fire blight orchard must be surrounded by a 500-metre buffer zone.

[4] Assessments provided included McKelvie *et al.* (1994).

- The orchard and buffer zone must be formally inspected at least three times a year by the US officials and the US/Japanese officials at harvest time.
- Harvested apples must be treated with sodium hypochlorite solution to disinfect them.
- Harvesting containers and the interior of the packing facility must be disinfected with chlorine.
- Fruit to be exported to Japan must be kept separate from other fruits post harvest.
- The US plant protection officials must certify the fruit free of fire blight and that appropriate treatments (above) have occurred post harvest.
- Japanese officials must confirm the US certifications and also inspect packaging facilities.

As Japan is a fire blight-free country and would be particularly sensitive to any infection from fire blight, it considered its strict phytosanitary measures to be justifiable.

3.4.2. *Use of SPS and other relevant rules*

The USA brought forward the dispute to the WTO claiming that Japan was violating WTO terms including Article XI of the GATT (1994) (elimination of quantitative restrictions), Article 14 of the Agreement on Agriculture (agreement to give effect to the SPS for agricultural products) as well as numerous articles within the SPS (Articles 2.2, 2.3, 5.1, 5.2, 5.3, 5.6, 6.1, 6.2 and 7 and Annex B of the SPS Agreement). These articles related to the following issues:

- whether the measures were necessary to protect life and health (or whether less stringent measures could be used);
- whether there was sufficient scientific evidence to justify the restriction;
- whether the measures by Japan arbitrarily or unjustifiably discriminate between members;
- the legitimacy of any risk assessments that have been undertaken, whether sufficient scientific evidence is considered and whether relevant economic factors are considered in any decision, and whether there are others — less trade-restrictive options — that meet the objectives of disease protection (Article 5 of the SPS);

- whether Japan had sufficiently considered whether its measures are appropriate to the context of the issue, for example, the level of prevalence of disease, the existence of existing disease eradication programmes.

3.4.3. *Use of data*

The Panel Report refers to the relevance of International pest risk management measures (ISPM 2). These state that the efficacy and impact of the various options in reducing risk to an acceptable level should be evaluated in terms of the following factors:

- biological effectiveness;
- cost/benefit of implementation;
- impact on existing regulations;
- commercial impact;
- social impact;
- phytosanitary policy considerations;
- time to implement a new regulation;
- efficacy of option against other quarantine pests;
- environmental impact.

It also states that 'zero risk' is not a legitimate option in any trade restriction and that the risk should be managed by considering interventions. Within the ensuing Panel debate, significant levels of data are provided on the scientific aspects of the case to support or discredit the disputed arrangement, including the following:

- application to the particular consignment;
- measures applied to prevent or reduce original infestation in the crop;
- measures to ensure the area or place or site of production or crop is free from the pest;
- consideration of other types of pathways (such as to curb natural spread);
- measures within the importing country;
- phytosanitary certificates and other compliance measures.

Within this case, Japan and the USA both provided vast amounts of scientific data surrounding the likelihood of fire blight arising in this case and the potential risk of changing Japan's ban. The USA contended that the trade was in mature apples where fire blight had never been an issue; Japan countered by noting 'hypothetical risk' of fire blight entering the supply chain though. Both parties held different positions on the acceptability of the risk assessment used. They highlighted a distinction between direct and indirect scientific evidence as part of this. The US continually contended that mature apples were zero risk and Japan repeatedly accepted this but insisted that the risk came from cross-contamination, processes used for packing, transporting fruit, contamination from nearby crops as a tangible risk of transmission pathways for the disease to get to Japan. Scientific data were further provided to support (Japan) or dismiss (USA) the nine restrictions/requirements for trade set out in the above introduction.

Although there was discussion on whether there are less trade-restrictive measures Japan could put in place, discussion in the report reiterates the issue of risk and scientific certainty (or otherwise) of disease transmission. No data are provided on the economic implications of the trade restriction or alternative measures to the disputed ban. Additionally, responses from an expert Panel are provided, which answered a range of questions relating to fire blight, its risk of transmission and spread, and circumstances that may aid or prevent it. The questions and responses are based on scientific data alone with no questions on the trade or economic impact of the measures.

As such, although trade restrictiveness is part of the discussion and the reference to the ISPM 2 highlights the legitimacy of considering cost/benefit of options, commercial and social impact, there are no data presented to this effect in this case.

3.4.4. *Outcome*

The original Panel and the Appellate Panel both found arguments in favour of the US. They determined the following:

- Under SPS Article 2.2, the measure by Japan did not present sufficient scientific evidence to justify it and that Japan's import ban was disproportionate for the negligible risk of fire blight.
- The risk assessment requirement (SPS Article 5.1) was not applied adequately for WTO requirements as it failed to evaluate (i) the likelihood of entry, establishment or spread of fire blight specifically through apple fruit and (ii) the likelihood of entry 'according to the SPS measures that might be applied' (WTO Appellate Report).

Additionally, the original Panel Report concluded that Japan had acted inconsistently with Article 5.6 of the SPS as the measure was more trade-restrictive than that required to achieve their appropriate level of sanitary or phytosanitary protection. It determined that the US alternative that apples imported into Japan should be mature and symptomless met the requirements of least trade restrictiveness as this alternative measure "(i) was reasonably available taking into account technical and economic feasibility; (ii) achieved Japan's appropriate level of sanitary or phytosanitary protection; and (iii) was significantly less restrictive to trade than the SPS measure at issue, and thus satisfied the three-pronged test confirmed by the Appellate Body in *Australia — Salmon*." (WTO Panel Report).

3.5. DS367: Australia vs. New Zealand — Measures Affecting the Importation of Apples from New Zealand (2007)

3.5.1. *Introduction*

In 2007, New Zealand brought a complaint to the WTO concerning the restrictions on imports of their apples into Australia. New Zealand considered that Australia's policy on imported apples, based on the Quarantine Act 1908 and its application of phytosanitary measures as specified in its Final IRA Report was contrary to the articles of the WTO in relation to least trade restrictiveness and its SPS.

Australia had adopted 16 phytosanitary measures for the import of apples from New Zealand, including eight measures adopted against the risk of fire blight, four against European canker, one against apple leaf curling midge (ALCM) and three applying generally.

New Zealand claimed that Australia's risk assessment also ignored available scientific evidence, Australian border inspection processes, relevant apple production processes in New Zealand, relevant diseases or pests in New Zealand and relevant climatic conditions in both New Zealand and Australia.

3.5.2. *Use of SPS and other relevant rules*

New Zealand contended violation of the WTO rules under the Articles 2.1, 2.2, 5.1, 5.2, 5.3, 5.5, 5.6, 8 and Annex C(1)(a). Between them, New Zealand alleged that Australian measures as follows:

- were maintained without scientific evidence;
- were not based on a proper risk assessment;
- subjected imported fruit with a degree of risk equivalent to or higher than that of New Zealand apples to measures substantially less restrictive than those on imports of New Zealand apples (discrimination);
- were more trade-restrictive than necessary to achieve Australia's appropriate level of protection (ALOP).

The issues under consideration were very similar to those in the Japanese apples case previously in that they were based on contesting the validity of the risk assessment and of the scientific evidence.

The notion of least trade restrictiveness was debated extensively in both the original Panel and the Appellate Board. This measure relates to Article 5.6 of the SPS which New Zealand contended were more trade-restrictive than required. The Panel noted the 'three-pronged approach' to evaluating any alternative measure as follows:

- being reasonably available taking into account technical and economic feasibility;
- achieving the importing member's appropriate levels of sanitary of phytosanitary protection;
- significantly less restrictive to trade than the SPS measure(s) at issue in the dispute.

In this case, New Zealand contended that either (i) the import of mature, symptomless apples and/or (ii) sampling inspection at import met these criteria, although no detail or discussion was provided as to the economic feasibility of either approach.

3.5.3. *Use of data*

As with other biosecurity cases, substantial amounts of scientific evidence (data) were provided on transmission and spread of disease, and extensive use of expert evidence was provided and referenced back to expert evidence in the Japanese apples case.

In terms of economic or trade data, this case hinted at its relevance. For example, under its discussion in the Appellate Body on risk assessment, it notes "the Panel found that the IRA had a tendency to overestimate the consequence of fire blight, particularly in respect of the criteria concerning plant life or health and domestic trade or industry..." But no data on such are provided.

Additionally, under Article 5.6, there was opportunity to provide data on the 'technical and economic feasibility' of alternatives to an import ban, but these were not provided as part of the debate or deliberations.

3.5.4. *Outcome*

The Panel reported in early 2010, but there was a subsequent appeal by Australia, with the Appellate Panel reporting in November 2010. In both Panels, the ruling overall was made in favour of New Zealand, with a requirement for Australia to put in place WTO-complaint measures, although New Zealand could not determine that Australia was non-complaint with all articles of the SPS set out by them. In more detail, the Panel found the following:

- The Panel found that specific measures regarding each of the three pests at issue, as well as the 'general' measures relating to these three pests, were inconsistent with Articles 5.1 and 5.2, and that, by implication, these measures were also inconsistent with Article 2.2 of the SPS Agreement, ruling against Australia. Australia appealed these findings only in regard to two of the three pests (fire blight and ALCM). The Appellate Body upheld the Panel's above findings regarding the two pests and the general measures relating to these two pests.
- The Panel found that New Zealand could not demonstrate that the measures at issue were inconsistent with Article 5.5 and, consequently, also could not demonstrate inconsistency with Article 2.3 of the SPS Agreement (prohibition on discrimination and disguised restriction on international trade).

The Panel also found that Australia's measures relating specifically to the three pests at issue were inconsistent with Article 5.6 (appropriate level of protection — alternative measures), but that New Zealand could not demonstrate that the three 'general' measures were inconsistent with Article 5.6. (Australia appealed these findings only in regard to two of the three pests (fire blight and ALCM). The Appellate Body reversed the Panel's findings of inconsistency in regard to the measures relating to these two pests but was unable to complete the legal analysis of New Zealand's claim.)

3.6. DS430: India vs. USA — Measures Concerning the Importation of Certain Agricultural Products (2012)

3.6.1. *Introduction*

In this case, the USA brought proceedings to the WTO against India for their import prohibitions on products purportedly because of their concerns relating to Avian Influenza (AI). India claimed it was acting in accordance with its national regulation, including the Indian Livestock Importation Act (1898) and its antecedents and a number of orders issued by India's Department of Animal Husbandry, Dairying and Fisheries.

AI, ('avian flu' or 'bird flu'), is recognised by the World Health Organization as "an infectious viral disease of birds (especially wild water fowl such as ducks and geese), often causing no apparent signs of illness". AI viruses can sometimes spread to domestic poultry and cause large-scale outbreaks of serious disease. Some of these AI viruses have also been reported to cross the species barrier and cause disease or subclinical infections in humans and other mammals (although transmission to humans is extremely rare).

AI viruses are transmitted by direct or indirect contact. Wild birds may cause the introduction of AI viruses in domestic poultry. However, according to the FAO, it is "unlikely that wild birds play a major role in spreading [AI] in poultry after its initial introduction" (although the Asian HPAI virus can be an exception to this) and that its spread is more likely within domestic poultry flocks where production and distribution practices may affect it.

The virus is known to have occurred in cats, dogs, pigs and related wild relatives and it is thought that in these cases the disease is transmitted through eating raw infected bird meat. Additionally, although unlikely, a small number of AI viruses can infect humans, with the impact of the HPAI virus having more serious consequences, including a small risk of fatality.

The Panel Report noted that "*One of the most well-known examples of transmission to humans is the HPAI subtype H5N1 virus present in poultry in certain parts of Asia and northeast Africa, which has caused human disease and deaths since 1997*". Within this case, India considered the impact of such an event could result in high levels of human health and mortality, even though the likelihood of the risk was low, and this concern was the driver for its import ban.

The Panel reported in 2014 and, after an appeal by India to its outcomes, an Appellate Body was formed and reported in 2015.

3.6.2. *Use of SPS and other relevant rules*

The USA considered that India's actions were in violation with numerous WTO articles, including the following:

- Articles I and XI of the GATT (1994) (elimination of quantitative restrictions);
- SPS Article 2 (basic rights and obligations 2.2 and 2.3);
- SPS Article 3.1 (harmonisation);
- SPS Article 5 (assessment of risk and determination of appropriate level of sanitary or phytosanitary protection — 5.1, 5.2, 5.5, 5.6 and 5.7);
- SPS Article 6 (adaptation to regional conditions — 6.1 and 6.2);
- SPS Article 7 (transparency);
- SPS Annex B (publication, notification and transparency requirements).

The US contended that India's measures were not based on relevant international standards (OIE Terrestrial Code) nor on an appropriate scientific risk assessment. In particular, they considered that India's import ban did violate the WTO requirement to only apply measures to the extent 'necessary' to protect human, animal or plant life or health (determined by scientific principles). Additionally, they considered that the measures constituted arbitrary or unjustifiable discrimination and that the ALOP specified by India was arbitrary and unjustifiable. In addition, the US considered the measures to be more trade-restrictive than required to meet a (scientifically determined) reasonable level of protection.

India considered that its measures did comply with the relevant international standard (OIE Terrestrial Code) and as such, the measures were scientifically legitimate, and by implication 'necessary' and trade-restrictive appropriate. Additionally, they argued that the transmission and spread of the disease was a higher risk given the geographical and climatic conditions compared to temperate regions.

3.6.3. *Use of data*

As with other cases, substantial data were provided on the scientific evidence and risk assessment aspects of this case. Expert opinion was sought on the OIE Terrestrial Code as well as on aspects of AI outbreak and transmission parameters for both the HPAI and LPAI typologies. This data included extensive data to help understand the domestic production and marketing of poultry and the possible risks of disease transmission within it.

The discussion under Article 5.6 provides the greatest likelihood of presenting trade or economic data, it being the article relating to whether there are other, less trade-restrictive measures possible. Precedent from previous cases noted that this should be achieved by examining whether there is an alternative SPS measure that: (i) is reasonably available taking into account *technical and economic* feasibility, (ii) achieves the member's appropriate level of sanitary or phytosanitary protection, and (iii) is significantly less restrictive to trade than the SPS measure contested.

In this case, as with the previous cases, discussion under this article focuses on technical solutions with little or no recourse to economic or trade implications *per se*. In this case, there is a hint of 'cost' of a measure, however as India argued that requiring veterinary certificates as recommended in the OIE Terrestrial Code is not technically and economically feasible because India does not have the capacity to handle the volume of imports that will result if it does not restrict imports during an active outbreak. No data were provided on the cost to increase such capacity and this assertion was rebutted by the US.

3.6.4. *Outcome*

The Panel and Appellate Body both found in favour of the US as follows:

- Both agreed that India's AI measures are inconsistent with Articles 5.1 and 5.2 as they are not based on a risk assessment and as a consequence were considered inconsistent with Article 2.2. Further exploration by the Appellate Body considered India was consistent with Article 2.2, but this did not condone its actions under Article 5.
- Both determined that India was in violation of Articles 3.1 and 3.2 as India's AI measures are not based on nor conform to the relevant international standard (OIE Terrestrial Code).
- Both agreed that India's claim to specificity of treatment relating to regional conditions under Article 6 was invalid.
- The Appellate Body upheld the Panel view that the US had presented as reasonable, less trade-restrictive alternative to the import ban imposed by India, finding India in violation of Articles 5.6 and 2.2.

3.7. DS447: Argentina vs. USA — Measures Affecting the Importation of Animals, Meat and Other Animal Products from Argentina (2012)

3.7.1. *Introduction*

Argentina brought a request for dispute resolution to the WTO in 2012 based on US restrictions on the import of animals, meat and other animal products from Argentina. This specifically relates to restrictions on imports based on US concerns for the spread of foot and mouth disease (FMD) into its territory from imported products. The import prohibition was originally enacted following an outbreak of FMD in Northern Argentina in 2001 through the US Code of Federal Regulations and Argentina was seeking to gain reauthorisation to import fresh (chilled or frozen) beef from Northern Argentina, and all FMD-susceptible animals and animal products from Patagonia.

FMD is a highly contagious viral disease that primarily affects cloven-hoofed livestock and wildlife. While most adult animals affected survive, morbidity rates are high with significant pain and distress seen, as well as decreased milk yield, permanent hoof damage and chronic mastitis. The disease can be transmitted through beef, offal and hides from infected animals.

Specifically, Argentina raised three issues:

1. the import prohibition of fresh beef from Argentina pertaining to the US rule from the Animal and Plant Health Inspection Service (APHIS);
2. the failure to recognise that parts of Argentina's territory is free from FMD, such that there should be no restriction on imports on that basis;
3. 'undue delays' in recognising the health status of a region or in granting approval to export animals or animal products.

Although this case came to WTO in 2012 with the Panel Report, findings were presented finally in 2015.

3.7.2. *Use of SPS and other relevant rules*

Argentina claimed that the continuing import restriction by the US more than 10 years after the 2001 outbreak of FMD in Northern Argentina violated numerous articles of the WTO. These were as follows:

- Articles I and XI of GATT (1994) (most favoured nation and quantitative restrictions);
- SPS Article 1 (general provisions);
- SPS Article 2 (basic rights and obligations – 2.2 and 2.3);
- SPS Article 3 (harmonisation — 3.1 and 3.3);
- SPS Article 5 (assessment of risk and determination of appropriate level of sanitary or phytosanitary protection — 5.1, 5.2, 5.4, 5.6 and 5.7);
- SPS Article 10 (special and differential treatment);
- Additionally, regarding the Patagonia ban SPS Article 6 (6.1 and 6.2) (regional treatment).

Argentina argued that the ban lacked scientific justification and was maintained contrary to international standards, did not have a valid risk assessment and was more trade-restrictive than required to meet the US desired level of protection. Argentina also noted that the US had enacted undue delay in the approval procedures on Argentina's requests to authorise imports of fresh beef from Patagonia once it was determined to be FMD-free. It also claimed that the US had not made due provision to adapt its measures to pest- or disease-free areas. Finally, under SPS Article 10, it considered the US had not given Argentina special and differential treatment accorded to it as a developing country member.

The US counted that its measures were required to continue the 80-year-old record of the country being FMD-free and that US cattle are not vaccinated against FMD, so a disease outbreak would be catastrophic. It considered its actions to be prudent, but acceptable, within this context and that its decisions were based on scientific understanding and international (OIE) standards. It used Article 8 SPS (control, inspection and approval procedures) and GATT Article XX(b) as part of its defence (necessary to protect animal health and life).

3.7.3. *Use of data*

As with the other biosecurity cases, the Panel extensively considered scientific data or information. This considered understanding FMD, its symptoms, damage caused and transmission. Expert opinion and research were presented relating to risk assessment techniques, veterinary practices and surveillance in the context of FMD. The Panel also consulted with the OIE (recognised international standard) concerning the operation and interpretation of its Terrestrial Animal Health Code.

As with the other cases, there is discussion on the extent of trade restrictiveness and whether the measure at issue could be resolved in a less trade-restrictive way. Also, Article 5.6 of SPS calls for this issue to be considered with respect to 'technical and economic feasibility' but there are no economic or trade data specifics provided. The US does note the potential devastating impact of an FMD outbreak in its jurisdiction but does not quantify this as part of the dispute.

3.7.4. *Outcome*

The Panel determined in favour of Argentina overall, considering the US measures to be inconsistent with Articles 1, 2, 3, 5, 6 and 8 of the SPS (but not Article 10 where it determined Argentina did not make a *prima facie* case). Based on the SPS decision, the US was also determined to be in violation of Articles I and XI of the GATT.

3.8. DS475: Russian Federation vs. EU — Measures on the Importation of Live Pigs, Pork and Other Pig Products from the European Union (2014)

3.8.1. *Introduction*

Between January and September 2014, the Russian Federation imposed an import ban (or highly restrictive limitations) on pigs, pork and other pig products from the European Union because of concerns about the prevalence of African swine fever (ASF) there and the risk of its transmission into the agricultural system in Russia. Along with the EU-wide ban, it also imposed individual imports bans from four EU member states (Estonia, Latvia, Lithuania and Poland).

ASF is a highly contagious disease affecting domestic and wild pigs, warthogs, wild boar, etc. Severe cases of the disease can result in high fever, loss of appetite, respiratory diseases, diarrhoea, haemorrhages in the skin and internal organs. Mortality rates are high among infected animals. Transmission between animals can occur through direct contact between sick and health animals and the virus can remain infectious for 3–6 months in uncooked pork products. Additionally, the virus can be transmitted indirectly through feeding on food waste including infected meat, as well as through poor hygiene in the processing environment (skin, clothes, vehicles, etc.). There is no vaccine for ASF.

An issue between the two parties was the means by which each was seeking to limit transmission of the disease. Russia claimed that it had undertaken an extensive wild boar hunting programme in the hope of reducing transmission as historically ASF was endemic in wild boar populations. The nation asserted that the hunting programme was aiding the virus' eradication and it did not wish to reintroduce it via infected products from the EU. ASF in the EU had been largely eradicated prior to outbreaks in 2014.

3.8.2. *Use of SPS and other relevant rules*

The EU challenged Russia's import ban on the basis of inconsistency with Articles I, III and XI of GATT (1994) as well as the following WTO SPS articles:

- SPS Article 2 (basic rights and obligations — 2.2);
- SPS Article 3 (harmonisation — 3.1, 3.2 and 3.3);
- SPS Article 5 (assessment of risk and determination of appropriate level of sanitary or phytosanitary protection — 5.1, 5.2, 5.3, 5.4, 5.5, 5.6 and 5.7);
- SPS Article 6 (adaptation to regional conditions);
- SPS Article 7 (transparency, and Annex B);
- SPS Article 8 (approval procedures).

The EU contended that Russia's measures are not in conformity with the OIE Terrestrial Code on numerous grounds (geography, product scope). It further contended that as it does not conform to international standards, it needs to rely on a legitimate risk assessment as an alternative, which the EU contended Russia failed to do (Articles 5.1 and 5.2), nor did it provide sufficient scientific evidence to justify its measures. It further asserted that the measures taken by Russia were not 'necessary' in accordance with Article 2 of the SPS, and that other less-trade-restrictive options were available which would still meet Russia's acceptable level of protection.

Russia argued that the import ban conforms to the relevant international standards (OIE) and therefore was not inconsistent with requirements under the SPS, nor did it apply the international standards in a discriminatory manner.

3.8.3. *Use of data*

Information was provided on factual aspects of ASF in terms of its typology, virulence and transmission, as well as data relating to how

ASF is considered within the international OIE Terrestrial Code standard. Scientific reports from the OIE, FAO, academics and veterinarians were provided as part of the dispute. Interestingly, there were less data-contended aspects of ASF than seen for the previous biosecurity cases, rather there was discussion on the 'science' related to the role of international standards versus other forms of risk assessment, so this discussion was more driven by legal process than factual information.

Within this case, reference is made to the economic impact of the import ban under consideration of Articles 5.3 and 5.6. The Panel Report (p. 217–80) noted as follows:

> *In this respect, the European Union limited its argument to a single paragraph in its first written submission, where it states without further explanation that "[i]n assessing the risk to animal health and determining the measures to be applied for achieving the appropriate level of sanitary protection, Russia failed to take into account all relevant economic factors referred to in Article 5.3 of the SPS Agreement, including the relative cost-effectiveness of alternative approaches to limiting risks".*

No data on the economic or trade implications of this measure were provided, with the EU noting that this would have been possible through the provision of an appropriate risk assessment.

3.8.4. *Outcome*

The Panel reported its findings in August 2016, with the Russian Federation taking the matter to the Appellate Body, which reported in February 2017.

The Panel determined that the matter was relevant to be considered under SPS rules. It also considered that the bans from the three of the four named EU countries do not conform to the relevant international standards and are thus inconsistent with Russia's obligations to base its SPS measures on international standards. Additionally, the Panel determined that Russia's approval procedures (Article 8 and Annex C) were in violation of SPS as they required information that was not necessary for the procedure at issue and the procedure resulted in undue delay.

The Panel did accept that Russia conformed to Articles 6.2 and 6.3 of SPS as it accepted the concept of pest- or disease-free areas or areas of low pest or disease prevalence with respect to ASF. While the EU provided appropriate information to Russia on areas within EU territory which are free of ASF, they were unable to provide the necessary evidence to objectively demonstrate that there are areas within Latvia that are likely to remain ASF-free. However, as Russia did not perform a risk assessment on which it could base its evaluation, the EU-wide ban and the bans from the four named countries were found to be inconsistent with Article 6.1.

The Panel determined that Russia did not base the measures at issue on a risk assessment within the meaning of the SPS agreement, breaching Articles 5.1 and 5.2 and 2.2 of SPS. It further noted that, as an appropriate risk assessment had not taken place, Russia had been unable to take into account relevant economic factors (SPS Article 5.3) and its measures could not be shown to be least trade-restrictive, taking into account technical and economic factors (Article 5.6). The Panel noted that the complaining party bears the burden to adduce sufficient evidence and arguments to raise a *prima facie* case that the responding party did not take into consideration the relevant economic factors listed in Article 5.3. It considered that the EU had not met this burden with respect to its claim that Russia did not consider those relevant economic factors, including the relative cost effectiveness of alternative approaches to limiting risks.

The Appellate Body confirmed the findings of the original Panel, with the exception of Article 6.2, as Russia recognised the concepts of pest- or disease-free areas and areas of low prevalence with respect to ASF. However, this alternation in the findings did not change the overall ruling in favour of the EU.

3.9. Summary of Analysis

These six biosecurity-related cases demonstrate how the SPS of the WTO has developed since its inception in 1995, as precedents become established for a range of decisions, including the following:

- the relationship between international standards on biosecurity (particularly the OIE Terrestrial Code) and SPS measures;
- the expectation of an acceptable risk assessment;
- the role of expertise brought in to the dispute in terms of academic credentials and the range of queries to comment on;
- the understanding of terms such as 'necessary' and the expectation on adaptation to regional conditions.

Each of the cases reviewed provided extensive data, information, fact and 'expert opinion' on the scientific aspects of biosecurity. This included discussions on the provenance, impact and transmission characteristics of the disease in question, including whether the disease impacted wild and domestic fauna.

The SPS rules imply the need to take account of relevant economic and trade factors within its consideration of disputes. It was therefore surprising to find that in no case was data provided on the economic impact of any disease outbreak in the country contesting the importation, its likely impact on the agricultural industry in terms of supply, or consumer price/gains from trade, or its impact on trade (volume or value). Given Articles 5.3 and 5.6, both are explicit in calling for consideration of 'technical and economic' circumstances and impact, and the need for a risk assessment (Articles 2 and 5) no data were forthcoming. As such, decisions were made on the basis of legal compliance with the evolving precedents in SPS rules, rather than the economic and/or trade impact of either party in the dispute.

3.10. Limitations of the WTO's Approach to Date and Alternative Options

There are a number of limitations to the approach used by the WTO to date in resolving biosecurity-related trade disputes, despite the introduction of scientific data and 'risk assessments' as part of the SPS. The core of the problem is the potential inability for a rules-based approach (using GATT (1994) and SPS articles) to deliver against organisational objectives. The WTO's objectives are to maximise social welfare through the mechanism of least trade restrictiveness, but the rules as they stand, and their interpretation through the dispute resolution process, has focussed on least trade restrictiveness in the context of a rules-based approach, with the addition of relevant scientific evidence. However, little regard is given to the trade and economic impacts of the dispute — and particularly no consideration is given to the likely consumer gains from trade. The lack of such data and its analysis results in Dispute Panels making decisions with only partial factual data available to them: they may know the scientific view of transmission and risk of spread of a non-native species or pathogen, but do not have the information to make an informed decision as to the trade, economic and social welfare impacts on either the exporting or importing country and its citizens.

Despite these limitations, the WTO has continued to approach trade–biosecurity disputes through interpretation of the SPS with scant attention given to economic, environmental or social impacts of trade. We contend here that in order to understand the holistic dispute under consideration, comprehensive data analysis of such impacts is required, including at the very least an evaluation of the likely consumer gains from trade. There are many advantages to such an approach. First, it widens the scope of what is considered relevant to the decision and as such increases the chance that the Panel will determine an outcome in line with WTO objectives. Second, it enables quantification and qualification of the facts to be considered and therefore the ability to assess relatively between impacts.

Previously, there have been practical barriers to implementing this approach, not least the ability to obtain meaningful data, to assess it based on rigorous valuation techniques and in determining acceptable levels of uncertainty (Haddock, 2000). However, following on from such limitations being identified, Part B of this book demonstrates how a data-led approach to analysing biosecurity-related trade disputes is now both possible and beneficial to social welfare.

Chapter 4

WTO Dispute Resolution Procedures in the Context of Ecosystem Services Risks — Details and Critical Analysis of Applications to Global Trade Disputes

4.1. Introduction

In Chapter 2, we introduced the key articles (or rules) used by the WTO in dispute settlement for cases relating to biosecurity and ecosystem services. How these are operationalised and their limitations were discussed for biosecurity-related cases in Chapter 3. In this chapter, we explore and evaluate ecosystem services-related cases.

For the purposes of this evaluation, we have used the definition of 'ecosystem services' from the Millennium Ecosystem Assessment:

> *the benefits people obtain from ecosystems. These include provisioning services such as food and water; regulating services such as regulation of floods, drought, land degradation, and disease; supporting services such as soil formation and nutrient cycling; and cultural services such as recreational, spiritual, religious and other nonmaterial benefits.*
>
> (Millennium Ecosystem Assessment, *Ecosystems and Human Well-being: A Framework for Assessment* (Island Press, 2003, pp. 1–25))

Within this chapter, we concentrate on ecosystem services other than those that directly provide services relating to food and water which have been dealt with in the context of biosecurity. Rather, we concentrate here on those aspects of ecosystem services which have value to people but which are not traded directly through markets. As can be deduced from the above definition, these may relate to protection/conservation of resources (such as land, air, water), to the maintenance of biodiversity and to cultural services, such as recreation and human health.

In this chapter, we discuss on a case-by-case basis how the GATT/WTO has handled each ecosystem services-related trade dispute. Each case is structured so that the first section summarises the case brought to dispute settlement, providing a brief description in relation to the ecosystem-services component (most cases have triggered the use and discussion of Article XX(b) and/or (g)). The second section provides an analysis of the core arguments used by the dispute Panel relating to this while the third section explores how 'data' or 'facts' relating to the impact of the situation under dispute have been used in the decision-making process. In the third section, such information is broadly categorised into trade, economic and environmental impacts, and the analysis highlights not only the extent of information considered but also whether it formed part of the WTO dispute Panel evaluation of the case. The final section provides the outcome from the dispute settlement Panel. This chapter concludes with a discussion of the processes and precedents used and the opportunities missed by omission of a data-centric evaluation component in the context of the WTO's stated aim of resolving disputes based on the objective of maximising social welfare.

4.2. Ecosystem Services Disputes

There has been a total of 12 disputes involving an ecosystem services feature since the incorporation of the GATT in 1948. The first case (Canada–US Tuna) was in 1982, and between then and the creation of the WTO, there were a further seven cases. Since the incorporation of the WTO in 1995, there have been four additional cases. These 12 cases are as follows:

- US–Canada, Tuna 1982 (L/5198);
- US–EC, Canada and Mexico Superfund Excise Taxes 1988 (L/6175);
- Canada–US, Unprocessed Herring and Salmon 1988 (L/6268);

- US–Thailand, Import and Taxes on Cigarettes 1990;
- US–Mexico, Tuna–Dolphin I 1991 (ILM 1594);
- US–Canada, Beer 1992 (DS17/R);
- US–Netherlands and EU, Tuna–Dolphin II, 1994 (DS29/2, DS29/3, ILM 839);
- US–EU, Taxes on Automobiles 1994 (Gas Guzzler and CAFE; ILM 1397);
- US–Venezuela and Others, Venezuelan Reformulated Gasoline 1996 (ILM 274);
- US vs. India, Pakistan, Malaysia and Thailand, Shrimp and Shrimp Products 1998 (DS58);
- Canada–EU, Asbestos and Asbestos-Containing Products 2001 (DS135);
- EU–Brazil, Retreaded Tyres 2007 (DS332).

4.3. US–Canada, Tuna 1982 (GATT Report L/5198)

4.3.1. *Introduction*

The US fishermen were found to be fishing for Albacore tuna within Canada's 200-mile Economic Exclusion Zone without authorisation from the Canadian Government and their boats were seized. US–Canada disputed jurisdiction and fishing rights. In retaliation, the US imposed an import ban on all tuna and related products from Canada. Canada brought this to GATT complaining of discrimination and a violation of the GATT. The main thrust of the dispute centred on Article XI, which argues that all trade barriers should be through tariffs only. Article XX was used by the US as a justification for their violation of Article XI, stating that it was a necessary non-tariff measure to protect an exhaustible natural resource.

4.3.2. *Use of Article XX and other relevant GATT Articles*

This case was a landmark case for later cases as it provided a precedent for how Article XX was to be interpreted. In particular, tuna was deemed to be an 'exhaustible natural resource'. The applicability of Article XX(g) was not accepted as a justification for the ban by the US though, as (i) could not be argued that the US tuna ban was primarily aimed at the conservation of natural resources, rather merely relating to it and (ii) whether it was considered necessary to use this means to protect the species.

4.3.3. *Use of data in dispute settlement*

In this case, some data were presented on relative trade, economic and environmental (tuna stock) impacts, but in no case were the data used as part of the decision process by the dispute Panel. In terms of trade and economic data, the US stated that its measures were not motivated by trade considerations, on the grounds of proportionality, as they only applied to trade valued at US$175,000 out of a total US import of tuna and tuna products of US$330 billion. In addition, the US provided data on production and consumption of tuna during the period. Canada noted its exports of tuna and tuna products averaged US$1.2 million.

Data on tuna conservation and species numbers were provided. The US noted that Albacore tuna, Yellowfin tuna and Bluefin tuna were all under intensive use and in some geographical areas in imminent data of depletion. Mention was made of international conventions on stock management signed by both jurisdictions in the dispute which apply catch limitations, although the US noted that Canada had, to date, refused to enter any international agreement on catch limits on Albacore tuna (the only species to enter its 200-mile fishery zone on the west coast).

4.3.4. *Outcome*

Canada and US had resolved the issue through bilateral agreement by the time the dispute Panel met, so no action was required by WTO. However, the Panel decided that the US action breached Article XI, so Article XX was the only means by which US action could be valid. The Panel determined some important precedents relating to interpretation of Article XX however. These were as follows:

- The assumption that tuna is an exhaustible natural resource was valid.
- That the acceptability of Article XX(g) and its provisions relating to the conservation of exhaustible natural resources was only applicable where the party concerned was also restricting domestic production and consumption to the same extent it was for imports. In this case, the US case was not accepted.
- The action of a trade embargo as a result of seizure of fishing vessels was not a justifiable action under the terms of Article XX.
- As such, the Panel concluded that, should this have needed settlement through GATT that the US did not provide sufficient evidence that the import prohibition complied with Article XX.

4.4. US–EC, Canada and Mexico Superfund Excise Taxes 1988 (L/6175)

4.4.1. *Introduction*

In 1986, the US congress mandated the funding of hazardous waste cleanup through the imposition of two new excise taxes (the 'Superfund Act'). This authorised a programme to clean up hazardous waste sites and deal with the public health problems caused by hazardous waste. The taxes proposed were as follows: (i) taxes on petroleum at 8.2 cents per barrel for domestic product and 11.7 cents for imported like products; (ii) the taxation of certain downstream chemicals equivalent to the tax they would have borne if their inputs had been sold in the USA; (iii) a further tax on certain imported substances sold or used by an importer, with the taxable substances being the derivatives of the chemicals subject to taxes in (ii) above. The imposition of the taxes was intended to remove an implicit subsidy on importers compared to tax liabilities of domestic producers.

The key areas for argument in this case were as follows:

- whether the petroleum tax differential between domestic and imported gasoline had a trade effect or not;
- whether the tax on certain imported substances was a case of nullification and impairment as it suggested that imported and domestic products were treated differently — the US arguing that domestic products were taxed during their production process so that this tax merely evened out the position;
- whether the measures were consistent with the polluter pays principle.

4.4.2. *Use of Article XX and other relevant GATT Articles*

This case did not use Article XX in its deliberations, but the argument centred on whether the notion of externalities/subsidies in different country production processes could be used to counter otherwise discriminatory behaviour (an exception to Article III:2) and where the cost of pollution should be borne — with the producer or user of the product.

There were two basic sub-cases within this dispute: petroleum tax differentials between imported and domestic products, and taxes on certain imported substances. For the former, the US argued that the differential did not have a significant trade effect and therefore did not violate GATT (Article XXIII), whereas the plaintiffs argued that even so, the action was still discriminatory under Article III. For the latter, taxes on certain imported substances, interesting environmental arguments were presented. For instance, the US claimed its actions were not discriminatory as they were applying taxes on importers that had been applied to domestic producers under the polluter pays principle. The plaintiffs countered that the tax was based on 'potential to pollute' rather than the pollution *per se*.

4.4.3. *Use of data in dispute settlement*

Detailed data of an economic nature, and its impact on potential trade, were provided by the US as part of their defence of the petroleum tax differential, although it was not used as part of the Panel's decision-making. They stated that the differential in tax between domestic and imported petroleum was so small that the commercial effects were insignificant — with a differential of US$0.0002 per litre being small compared to daily fluctuations in contract prices — so would not affect buying decisions or investment decisions to stimulate investment in domestic oil production. The US further provided information on oil market elasticity arguing the elasticity impact being –0.1 of this measure such that import levels would hardly be affected. The plaintiffs disagreed with these economic and trade impact claims, considering oil markets to be more elastic with price sensitive buyers, such that there would be an adverse impact on importers.

For the second sub-case, tax on certain imported substances, some environmental data were presented, but they were not used by the Panel as part of the decision-making. The US argued that in the case of petroleum, pollution occurs in its consumption as well as production, and as such, applying polluter pays means importers should pick up some taxation liability for the use of their product in the US. The plaintiffs argued against this, referencing OECD guidelines, noting that there was the potential for an importer to be paying twice through the tax systems in their own country (if PPP was applied) but also in the US, and also that the measure was being applied inconsistently with petroleum imports taxed even where it is used for the production of pollution-benign products (e.g. polythene and synthetic rubbers).

4.4.4. *Outcome*

The Panel concluded that tax on imported petroleum was inconsistent with Article III:2 (ruling against the US) as domestic and imported products were treated differently, regardless of the trade effect. For taxes on certain imported substances, the Panel concluded that it was a tax adjustment that was in principle consistent with Article II:2, and the existence of penalty provisions did not constitute an infringement as the tax authorities had regulatory power to eliminate the need for the imposition of the penalty rate, therefore the actions of the US were upheld.

4.5. Canada–US, Unprocessed Herring and Salmon 1988 (L/6268)

4.5.1. *Introduction*

The US complained that Canadian action under The Canadian Fisheries Act (1970), resulting in legislation whereby British Columbia were able to ban the export of unprocessed sockeye and pink salmon and all unprocessed herring, was contrary to GATT regulations. In this case, the US cited both Article XI export restrictions and Article XX(g). With regard to Article XI, Canada contended that its actions were permitted as they related to and were necessary to maintain standards and marketing of the fish which it considered to be a commodity. This was the first case where Article XX had a substantial airing in terms of decision-making (see Section 4.5.2) and set precedents for future cases in determining how the language of Article XX(g) should be interpreted. The main arguments addressed in the case were as follows:

- whether the export restrictions by Canada related to conservation were determined by the Panel to mean 'primarily aimed at';
- whether the intent of the ban was environmentally or economically motivated;
- whether the fish species were within the scope of 'exhaustible natural resources';
- whether the export measures were directly related to the alleged conservation objectives.

4.5.2. *Use of Article XX and other relevant GATT Articles*

This case looked to precedent from US–Canada Tuna 1982, concurring that fish were an exhaustible natural resource. The case also provided a valuable ruling on what the term 'relating to conservation' in Article XX(g) meant, determining that it means 'should be primarily aimed at conservation'.

Canada argued that as the measure did 'relate to conservation' in the clear sense of the term as they had a long-term and positive impact on international trade, as fish stocks could be protected. It further argued that Article XX(g) did not require the measure to be 'essential' or 'necessary'. The US countered by claiming that a measure that has a positive impact on conservation is not, in itself, sufficient justification for trade restrictiveness.

4.5.3. *Use of data in dispute settlement*

Some mention was made of economic and trade impact arising from this dispute. Canada noted that sockeye and pink salmon and herring fisheries represent the largest share of the West Coast Fisheries of Canada and they supply a dominant share of the processing sector, giving employment to five-sixth of the workers in the sector in British Columbia. The US argued that unrestricted exporting was economically desirable for both countries and would allow processors to extend their production runs and unit costs. This data was not used by the Panel in its consideration however.

There was some discussion around data relating to environmental conservation, but again the information was not considered relevant by the Panel in its ruling. Canada noted that conservation measures in their case had been deemed necessary as the species was becoming vulnerable, thus needing detailed reporting systems. Further, they noted the highly cyclical nature of both fisheries and the complexity of their management, the need to provide steady supplies of fish to the processing plant and the sensitivity of the species to quality control problems prior to and during processing. This measure would enable detailed and accurate statistics to ensure optimal production, and conservation, levels. The US did not disagree with the need for conservation measures but noted that these, export-restrictive, management practices were not required by other species with similar ecological conditions.

4.5.4. *Outcome*

The Panel found that Canada's actions could not be justified by Article XI as they were neither necessary nor relating to regulation for the marketing of fish. With regard to Article XX(g), the Panel interpreted that the intent was not to widen the scope for measures serving conservation but to ensure that GATT did not hinder their pursuit. As such, it determined that any trade restriction had to be primarily aimed at the conservation of an exhaustible natural resource to be legitimate. In this case, the Panel concluded that

Canada's measures had an effect on conservation but that was not their principal aim, such that the US complaint was upheld.

4.6. Thailand–US, Restrictions on the Importation of and Internal Taxes on Cigarettes (DS10/R)

4.6.1. *Introduction*

Under the 1966 Tobacco Act, Thailand prohibited the importation of cigarettes and other tobacco products, but authorised the sale of domestic cigarettes; moreover, cigarettes were subject to an excise tax, a business tax and a municipal tax.

The US complained that the import restrictions were inconsistent with GATT Article XI:1 (which banned quantitative prohibitions and restrictions on imports) and considered that they were justified neither by Article XI:2(c) (relating to whether tobacco was considered an agricultural product) nor by Article XX(b) (necessary for human, animal or plant life and health). It also argued that the internal taxes were inconsistent with GATT Article III:2.

Thailand argued that the import restrictions were justified under Article XX(b) as their government had adopted measures that were only effective if cigarette imports were prohibited. In this instance, they claimed that chemicals and other additives contained in US cigarettes might make them more harmful than Thai cigarettes and also that increased competition arising from the availability of imported cigarettes would lower prices and enhance availability which ran counter to their public health initiatives to curb smoking.

4.6.2. *Use of Article XX and other relevant GATT Articles*

Thailand contended that its curb on the import of cigarettes was justified by its public health policy objective to reduce the consumption of tobacco which was harmful to human health and was therefore justified under Article XX(b). It further contended, however, that it did not wish to prohibit tobacco production in Thailand because this could lead to the production and consumption of narcotic drugs (opium, marijuana, kratom) that have more harmful effects than tobacco. It further noted that domestic production of tobacco through the state monopoly was sufficient to meet, but not increase, domestic demand and, further, the state had put in place measures to discourage smoking (public notices, bans on advertising, etc.).

Finally, Thailand contended that imported cigarettes from the US brought additional health risks arising from the addition of various additives, over domestically produced cigarettes.

The US countered Thailand's contentions on the following bases:

- Using precedent from a previous GATT Panel, it noted that Article XX could only be legitimately implemented if an alternative measure which it could reasonably expected to employ and is not inconsistent with other GATT provisions is available to it. Note the precedent referred to Article XX(d) but the US contended consistency throughout Article XX sub-clauses. The US stated that Thailand could pursue the objective of preventing the increase in the number of smokers without imposing an import ban.
- The US further contended that Thailand's objectives to limit smoking was not working as sales were rising annually in the country. Thailand countered this by noting this related to increased population levels and a decreasing percentage of smokers within this.
- Thailand had done little to limit the availability of cigarettes, with over 40,000 cigarette retailers in the country and a variety of products to appeal to a range of market segments.
- The US cigarettes did not have special health concerns but offered a product with a lower nicotine and tar content than Thai domestic cigarettes. As such, there was no higher health risk from smoking imported cigarettes.

4.6.3. *Use of data in dispute settlement*

Within this dispute, significant data were provided, including a contribution from the World Health Organization (WHO) relating to the following:

- numbers of citizens smoking and percentage rates of smokers on a longitudinal basis;
- the chemical composition of Thai and imported cigarettes and the relative 'harmfulness';
- total domestic sales of cigarettes and share of the market by 'illegal' imports.

There was, further, discussion as to whether the availability of the cigarette market to importers and the increased competition would lead to an increase in overall consumption in Thailand. In addition, discussions occurred on the role of advertising, cheaper products, greater availability and products targeted at women.

Trade data were provided also on the increase in US exports of cigarettes to less developed markets including Asia and Latin America as the markets opened up and state monopolies were dissolved. This was coupled with data on smoking rates from the WHO which noted that per capita rates in the US were 3,200 cigarettes per person per year, whereas in Thailand, it was 900 in 1985 having reduced from 1,100 in the late 1970s.

The US further provided data to quell the Thai contention that it was attempting to restrain domestic production for health reasons by noting the expansion plans for production by the Thai Tobacco Monopoly between 1987 and 1990 to increase its production by 10 billion cigarettes by 1991.

4.6.4. *Outcome*

The Panel found that the import restrictions were inconsistent with Article XI:1 and not justified under Article X1:2(c) although the internal taxes were found to be consistent with Article III:2.

It further concluded that the import restrictions were not 'necessary' within the meaning of Article XX(b) and as such could not be condoned. In its findings against Article XX(b), it reported its views on health relating to both quality and quantity of cigarettes smoked as well as how cigarettes were supplied to the market and provided recommendations to Thailand of measures it should take that would be consistent with the GATT. In terms of the former, it noted other countries had introduced strict, non-discriminatory labelling and ingredient disclosure regulations which allowed governments to control, and the public to be informed of, the content of cigarettes. In terms of quantity smoked, the Panel recognised the benefits of reduced levels of smoking (and the WHO data and views) but stated that this could be achieved through a general ban on all cigarette advertising (both domestic and imported). In terms of how Thailand might restrict supply, the Panel considered the Thai government may use its Thai Tobacco Monopoly to regulate the overall supply of cigarettes, their prices and their retail availability produced, and it does not accord imported cigarettes less favourable treatment than domestic ones.

4.7. US–Mexico, Tuna–Dolphin I 1991 (ILM 1594)

4.7.1. *Introduction*

This case is the most well-known trade–environmental dispute and one that has created the greatest debate, as well as spawning a subsequent case (see later, Tuna–Dolphin II). The US imposed an embargo on the import of commercial yellowfin tuna and associated products harvested with purse-seine nets in the Eastern Tropical Pacific, until the country concerned could show that it was operating in compliance with the US Marine Mammals Protection Act (MMPA) 1972 aimed at protecting dolphin species. The Act states that importation could only be allowed if:

- the government of the harvesting country has a programme regulating the taking of dolphins that is comparable to that of the US;
- the average rate of incidental taking of dolphins by vessels of the harvesting nation is comparable to the average rate of taking by the US vessels. These rates for any country's fleets must not exceed 1.25 times the average taking rate of US vessels in the same period.

Additionally, the MMPA required the government of any intermediary nation from which yellowfin tuna or its products are imported to the US to certify and provide reasonable proof that it has acted to prohibit tuna importation from a nation from which direct export to the US is banned (the intermediary nation embargo).

Mexico complained to the GATT that the US import prohibitions were inconsistent with Article XI as they were quantitative restrictions and that the different treatment of domestic and foreign fishing violated Article III. The US argued that its tuna was not a 'like product' to Mexican tuna as it was fished using a different process and that 'purse-seine' net fishing allowed a higher kill rate, but the externality of dolphin loss. Article XX was discussed by both parties in terms of its legitimacy in this case as well as its scope (see Section 4.7.2).

4.7.2. *Use of Article XX and other relevant GATT Articles*

This case contains a range of significant legal arguments relating to Article XX(b) and (g). In summary, the core debates occurred around definitions of terms within the Article, as well as the jurisdiction it could operate in. Debate occurred around the term 'necessary' as with previous cases.

Additionally, there was discussion as to whether Article XX(g) related to individual dolphins or the dolphin population as a whole. This was particularly relevant as the US was arguing that it was the context-specific elements of the dolphin population that was relevant (where numbers of individuals were threatened to the point of being exhausted). Whether the population had to be 'exhausted' or 'exhaustible' (as discussed in above cases) was also debated in terms of the validity of applying Article XX(g). Substantially, this case introduced the notion of 'extraterritoriality' into GATT decision-making, with Mexico questioning whether the US could legitimately control assets (in this case dolphin and tuna) outside of their jurisdiction. It was noted that as marine species they did not necessarily locate within one territory, protecting them through international means was appropriate. It was also debated as to whether tuna caught through purse-seine nets were a different 'product' to line-caught tuna as the process of 'production' differed.

4.7.3. *Use of data in dispute settlement*

Within this case, there were no data relating to economic or trade impacts of this restriction on either party. A note was made about the geographical and temporal arrangements for coverage of the restrictions under the MMPA, for context, to show it was not directly discriminating against Mexico.

This case did provide extensive data and information regarding the environmental impacts of purse-seine fishing and dolphin species conservation. The US explained how the fishing process worked and how the requirements for tuna production needed to be modified to limit the kill of dolphin. It noted the technical measures it took to ensure this, including specific gear and fishing procedures, prohibition of explosive and speedboats, enhancement of management and performance standards, etc. Within this, there was an implication of a 'production cost' relative to the more 'cost effective' means through purse-seine fishing (as long as dolphin kill was not seen as a 'cost'). The US further provided extensive data on dolphin mortality rates as well as migratory behaviour (moving beyond specific jurisdictions), justifying its import restrictions under the MMPA, in the absence of suitable international agreements, e.g. the Convention on International Trade in Endangered Species of Wild Fauna and Flora (CITES).

Mexico noted that none of the species covered by this restriction were highlighted by CITES as in danger of extinction, nor were they taxonomically close to those covered by it. They disagreed with the US data on dolphin population depletion risks, noting that many populations have remained stable in the region. Finally, they noted that as the larger tuna swim with dolphins, there would be a lower catch of these (older) fish at the expense of younger ones, thus risking depletion of stocks as from the standpoint of both economics and conservation, it was more rational to use a renewable resource after it had reproduced rather than before.

Despite the extensive discussions within the dispute of dolphin taxonomy, conservation and behaviour, as well as implicit costs of fishing methods, this was not considered by the Panel in its ruling.

4.7.4. *Outcome*

The GATT found arguments against the US in this instance. The Panel found the US action to be contrary to Articles II, XI and XX of the GATT. With respect to Article XX, it determined the following:

- The process by which tuna was caught did not affect the product as such (i.e. that process and production method was not a legitimate means to distinguish between products).
- Article XX could not be applied extraterritorially (i.e. the US could not claim it wishes to protect the environment outside of its own jurisdiction as part of its trade restrictions).
- There were other means to conserve dolphins so the trade restrictions were not 'necessary'.
- The mechanisms of establishing whether Mexico had violated US requirements on dolphin catch could only be determined retrospectively and therefore was unviable.

4.8. US–Canada, Beer 1992 (DS17/R)

4.8.1. *Introduction*

In this dispute, the US brought a complaint to the GATT against the sales and marketing practices used to sell and distribute beer in Canada, claiming that domestic and imported beers were treated differently. Although this was a wider dispute, the environmental aspects of the case concerned Canada's contention that its tax measures on US beer bottles was necessary for environmental protection, although the US countered the measures were anti-environmental as the bottle collection system was set up in such a way that it was not viable for foreign producers to collect and recycle bottles.

4.8.2. *Use of Article XX and other relevant GATT Articles*

This case used Article XX(b) by Canada as a means to defend its use of a tax charged on beer bottles, designed to encourage the establishment of systems in which customers returned bottles for refilling and cans for recycling through a tax rebate to the customer on return recycle. The use of the tax was to offset (internalise) the wider costs of disposal to landfill.

However, the US argued the system was counter to GATT objectives as the tax was refundable on domestic beer containers as domestic producers could collect containers through private delivery systems they were allowed to operate, whereas foreign containers generally distributed through a state system which had no integrated collection service (and therefore had to pay the tax in lieu — beer sales were too small to justify commercially their own collection services). This meant the customer could not benefit from a tax rebate on container return in the same way, with the US contending that Canada's measures worked counter to environmental objectives.

4.8.3. *Use of data in dispute settlement*

Relatively little data were provided by the parties in this dispute, and none of it was used by the Panel as part of its decision-making. In terms of data relating to economic impact, the US noted that there was a direct economic impact on their brewery business as their beer was forced to be levied with a tax which domestic beer could avoid by the use of the returnable bottles scheme. Delivery could not be combined with collection and the cost of establishing a separate collection system was prohibitively expensive. In contrast, domestic brewers were able to collect used containers more economically as they could combine collection with delivery. To counter this, Canada argued that it should not be the responsibility of the Canadian government to pay the high costs of disposal and that this should be the responsibility of the manufacturer (i.e. the producer pays). It also noted that the decision of the Canadian breweries was a private commercial one.

With regard to environmental data, Canada noted the importance of the environmental issues in this case, with levies on cans and bottles as part of deposit and return schemes established to encourage recycling. Where no such scheme was established the charges were intended to offset the cost of disposal of the containers. This argument links with the data above and the principle of 'producer pays' for the polluting impact of a non-recyclable product.

4.8.4. *Outcome*

The Panel looked to Article III of the GATT in its decision. It determined that the environmental tax on beer containers were not in violation of Article III but their application in a situation where different systems for delivery of beer applied between domestic and imported beer. It determined that the differential processes between domestic and imported beer containers' collection were inconsistent with the GATT and by allowing foreign brewers the same distribution rights as domestic brewers, the problem of differential environmental taxing would be resolved.

4.9. US–Netherlands and EU, Tuna–Dolphin II 1994 (DS29/2, DS29/3, ILM 839)

4.9.1. *Introduction*

This case follows from the 1991 US–Mexico Tuna–Dolphin dispute and the acceptability of provisions in the US MMPA which enabled primary and intermediary nation embargoes of tuna products where there had been incidental dolphin kills beyond acceptable (by US measures) levels. In this case, the dispute centred on the intermediary nation embargo, as opposed to the case with Mexico, who were primary traders. The EU objected in this case to the action by the US whereby imports were banned from that nation unless it could provide certification and proof that it had prohibited the import of any tuna that had been barred from entry into the US.

4.9.2. *Use of Article XX and other relevant GATT Articles*

This case raised the use of Article XX(b) and (g), debating again whether the measures were 'necessary' for protection (i.e. least inconsistent with the GATT) and primarily aimed at conservation. However, in this case, two additional arguments were raised. The first related to the notion of 'extraterritoriality', i.e. whether the US could impose its laws on other nations. This had been discussed in the first Dolphin–Tuna case but was expanded here. Second was the issue of what the relationship was between the GATT and multilateral environmental agreements or related treaties. The Panel concluded, in relation to both of these, that Article XX did not support the US position. It recognised in this dispute the validity of US conservation objectives was not in dispute, rather whether it could impose trade embargoes to secure changes in the policies which other contracting parties pursued within their own jurisdiction in pursuit of US environmental objectives.

4.9.3. *Use of data in dispute settlement*

Extensive environmental data were presented in this dispute, although it was not used by the Panel as part of its decision-making. No trade or economic data were provided or considered.

With regard to environmental information, the plaintiffs (Netherlands, EU) provided data on the number of dolphin kills in the disputed area (Eastern Tropical Pacific or ETP) in the 1970s and 1980s, noting dolphin numbers were reducing. They further provided data on dolphin kills by purse-seine nets, noting a reduction from 1,33,000 in 1986 to 27,500 in 1991 with the reduction representing a growing consensus on the need to reduce mortality as well as the introduction of improved fishing methods and equipment. At the time of the dispute, dolphin mortality levels were stable (about 15,500), representing less than 1% of the population in the area.

The US noted this, but cited their intent, through the agreement of IATTC members to reduce mortality to less than 5,000 by 1990 and to find an ecologically sound means of harvesting tuna without affecting dolphins. The US further argued for the need, through the GATT, to recognise methods of protecting the environment, including the global commons, and as such, the extraterritorial nature of their measure was valid. As such, it called for the GATT to enable a party to mandate for environmental conservation reasons beyond its jurisdiction.

4.9.4. *Outcome*

The Panel noted the objective of environmental sustainability by the US but observed that the issue in this dispute was not the validity of US conservation objectives but rather whether it could impose trade embargoes to secure changes in the policies which other contracting parties pursued within their own jurisdiction in pursuit of US environmental objectives.

The Panel determined that none of the issues in this dispute could be represented legitimately through the application of Article XX and as such found in favour of the plaintiffs.

4.10. US–EU, Taxes on Automobiles 1994 (Gas Guzzler and CAFE; ILM 1397)

4.10.1. *Introduction*

This case dealt with two main disputes: a gas guzzler tax on all cars with fuel economy below 22.5 mpg and a corporate average fuel economy charge (CAFE) based on the corporate average fuel economy for a manufacturer. The EU claimed with the gas guzzler tax that it was not primarily aimed at conservation in the clear sense, as such measures were not applied to many fuel economy devices (e.g. trucks, power stations) and that the tax was discriminatory as it largely impacted foreign cars (e.g. BMW, Mercedes). For CAFE, the US contended that the system was not discriminatory, rather measures impacted different manufacturers depending on their market position/conditions not country of production/export *per se*. The key areas of argument in the case were as follows:

- whether all cars, regardless of their specification, could be considered to be 'like products';
- whether the tax could be considered to be primarily aimed at conservation;
- whether the tax has had an unjustifiable detrimental effect on European car sales to the US;
- whether the tax was the most appropriate means of meeting energy conservation goals.

4.10.2. *Use of Article XX and other relevant GATT Articles*

Article XX(g) was used as part of the debate. First, there was discussion as to proportionality with the EU contesting that as EU imports made up less than 4% of the car market in the US, taxes on these vehicles would have a relatively insignificant environmental impact. The US contended the taxes were part of incentivising fuel-efficient behaviour and thus relevant to Article XX(g) and conservation, although the EU contested its significance given other vehicles, e.g. light trucks were excluded. Discussion took place as to whether it was valid to consider a measure which indirectly impacted conservation of natural resources or whether Article XX required direct impact only.

4.10.3. *Use of data in dispute settlement*

This was the first case where the use of data become more prominent, with some trade data being used by the Panel in its ruling, although the extensive environmental data provided were not used. The data used are presented separately for the gas guzzler and CAFE disputes separately below.

4.10.3.1. *Gas guzzler*

The EU provided details of the scale of car production and trade between the US and EU, noting the lack of equity as 0.5% of US cars were subject to the tax whereas 18% of European cars imported to the US were affected by it, and that in 1990, 73.36% of all gas guzzler revenues were derived from European manufactured cars. They provided further details on these trends over time, showing a consistent pattern, and noted that this had led to a decline in sales of large expensive cars, which made up many of the imports to the US. The US countered this by providing data suggesting EU manufacturers had the option to produce more fuel-efficient cars (as the US manufacturers were doing) but were choosing not to do so. It further countered EU claims that fuel inefficiency was a necessary attribute of luxury class vehicles by citing performance characteristics of the Japanese luxury brand, Lexus.

The EU considered the gas guzzler tax had upset trade relationships with the US as it was giving vehicles not subject to the tax a competitive advantage *per se*. Also, the system of recording performance was based on an average of all cars sold by the manufacturer in the US, so that US manufacturers serving the full market spectrum could average their results, hiding poor performance of its luxury marques. The EU could not do this, as the market for EU vehicles sold in the US were at the luxury end (although in reality, BMW and Mercedes at the time did not offer more compact mass market vehicles).

The US cited data showing that the tax had been successful as a fuel conservation (and therefore environmental) measure as it had created incentives to shift manufacturing and consumer purchasing to more fuel-efficient vehicles. It noted that in 1978 over 75% of passenger vehicles would have been subject to the tax (as their fuel efficiency was below 22.5 mpg), less than 1% of vehicles failed this at the time of the dispute. The EU countered the relevance of this data as a means of fuel conservation as follows:

- The threshold affected just 0.7% of vehicles sold, 85% of which were imported. If fuel conservation was a legitimate aim, the fuel efficiency stipulation would be higher.
- US manufactured vehicles with lower fuel efficiency than this were escaping the tax because of the calculation methodology.
- The tax was not imposed on minivans, pickup trucks or sport utility vehicles (SUVs) ever though their fuel efficiency was low.

4.10.3.2. *Corporate average fuel economy (CAFE)*

As with the Gas Guzzler tax, the EU argued about the differential impact on imported vehicles, noting that since the beginning of CAFE no vehicle produced by the big three US car manufacturers had been burdened by it (with total payments between 1980 and 1992 of US$263 million, 99.9% from EU cars). They further noted that the US international car market tended to be more specialised than the domestic market, with EU models appealing to the luxury/high end which tended to be less fuel efficient. The US countered these arguments by stating, as with gas guzzler, that the EU manufacturers had chosen to manufacture and sell less fuel-efficient cars and that domestic manufacturers would be unfairly penalised if this regulation was not in place (as they had effectively turned away from selling more powerful, fuel inefficient vehicles). They further noted that the disproportionate impact on the EU ignored the compliance costs that US manufacturers had borne in compliance. They provided extensive details on how car specifications had changed in the US since the regulation and the costs to US manufacturers (e.g. it noted Ford spent $12 billion between 1978 and 1986 and it intended to spend another $3 billion over the following 2 years on fuel economy programmes). The US further argued that taxing fuel itself would be regressive as it would not encourage technological innovation that ensures fuel conservation in the long term but would also adversely impact rural communities and lower income individuals.

Environmentally, the US stated that the CAFE approach offered a dramatic decrease in fuel consumption compared to any other means of fuel conservation, providing data showing how the average fuel consumption had improved to 28.3 mpg compared to 14 mpg in 1974 and that total fuel consumption in this period had declined by 11% at a time when total auto registrations had increased by 23% and total miles driven increased by 32%. They further noted that petroleum as a fossil fuel is an exhaustible natural resource with adverse impacts on climate change from its use. It noted that, as the US transport sector comprised 7% of global emissions finding a means for their reduction was vital. To counter these arguments, the EU argued for the use, instead, of a non-discriminatory tax on fuel consumption as all vehicles with the same fuel consumption (regardless of provenance) would face the same disincentive. Additionally, it may disincentivise use of the vehicle, thus reducing total miles driven in a manner that a tax on car purchase, as a one-off cost regardless of use, would not.

4.10.4. *Outcome*

The use of Article XX was central to the Panel's decision-making. In considering the article, the Panel suggested a three-step analysis:

- whether the policy fell within the scope of exhaustible natural resources;
- whether the exception was 'related to' the conservation of exhaustible natural resources and whether it was made effective 'in conjunction with' restrictions on domestic production or consumption;
- whether the measure did or did not arbitrarily or unjustifiably discriminate between countries, or whether it was a disguised restriction on international trade.

The Panel agreed gasoline was an exhaustible natural resource, but that the US provided no evidence that not allowing fuel efficiency levels of large imported cars to be merged with small domestic cars under the same ownership or control of production furthered the objectives of resource conservation. Rather it suggested that the foreign fleet accounting method served to inhibit imports of small cars and so was not justified under Article XX.

However, the application of the fleet averaging system in a manner similar to that of domestic cars clearly served the purpose of fuel conservation and that if there was no requirement placed on imported cars at all, the objectives of the CAFÉ programme would be prejudiced in favour of large imported cars which would have no restriction on fuel consumption. As such, the fleet averaging method overall could be justified under Article XX.

4.11. US–Venezuela and Others, Venezuelan Reformulated Gasoline 1996 (ILM 274)

4.11.1. *Introduction*

This was the first case to be dealt with under the guise of the WTO instead of GATT. The arguments and style of dispute settlement are much the same as previously though, with the exception of discussion as to whether actions conformed with Technical Barriers to Trade (TBT). The case centred around different requirements between domestic and foreign gasoline suppliers, in this case for 'reformulated' gasoline, the only gasoline product that could be sold in designated poor air quality parts of the US (accounting for 30% of gasoline marketed in the country). The US EPA decided to create different rules for the establishment of baselines, depending on whether the manufacturer was domestic or not. For the former they were allowed to establish their baseline by one of three methods. A foreign refiner could only establish their own baseline if it imported at least 75% (by volume) of its gasoline into the US, otherwise it had to comply with the statutory baseline. The key arguments in this case centred around the following:

- differential treatment between foreign and domestic manufacturers;
- whether the measures by the US were 'necessary' by definitions within Article XX;
- that the measures on importers were necessary as US legislation could not be enforced outside of the US;
- whether the differential standards for domestic and imported gasoline were relevant given the small percentage levels of imported gasoline compared to that domestically produced;
- whether clean air could be considered to be an exhaustible natural resource;
- whether the gasoline rule had had a significantly detrimental trade and economic effect in Venezuela and Brazil.

4.11.2. *Use of Article XX and other relevant GATT Articles*

The US contended that their actions fell within the scope of Article XX(b), (d) and (g) and was consistent with the Agreement on TBT.

For Article XX(b), the US noted that the measure was 'necessary' as toxic air pollutants impacted human health and ozone was responsible for agricultural crop loss. Venezuela/Brazil considered the measure not 'necessary' as there were other less trade-restrictive measures that could still meet environmental objectives that could be used. The US countered that the statutory baseline was fair and a means to ensure that standards of imported products met US requirements, in a situation where law enforcement was not possible outside of US jurisdiction (the issue of extraterritoriality as in Tuna–Dolphin I). The plaintiffs then queried why domestic producers could not use the statutory baseline also.

Venezuela/Brazil also referenced the notion of proportionality and the use of a trade-restrictive measure (for environmental reasons) when the imported product made up 2% of total market, thus the environmental benefit was disproportionate to the trade cost.

The US argued under Article XX(g) that clean air was an exhaustible natural resource as it could be exhausted by emissions of pollutants, such as VOCs, NO_x and toxins. It considered clean air was similar in concept to fish (precedent from case 3 Herring–Salmon) as both could be depleted or rendered extinct. It also noted the restrictive nature of the gasoline rule on domestic production by requiring domestic producers to limit their production too.

The plaintiffs argued against the validity of Article XX(g) as clean air was a 'condition' of the air that was renewable rather than a resource that was exhaustible. It noted, with precedent from case 1 (Canadian Tuna), that the GATT had interpreted the meaning of 'exhaustible' narrowly and from this clean air should not be considered an exhaustible natural resource. Further, it noted that the precedent of 'relating to' had been interpreted as 'primarily aimed at'. However, if the gasoline rule was primarily aimed at conservation in this case, its methodology contained loopholes that undermined its conservation objectives, not least as the restrictions related to quality, not quantity, of emissions.

With regard to the TBT, the plaintiffs considered that the gasoline rule was within its scope for consideration by GATT/WTO, as it was a document that laid down product characteristics with which compliance was mandatory and under this was in violation with Article 2.2 as follows:

- The US had prepared the rule with an intent to create obstacles to international trade.

- The more stringent requirements on foreign manufacturers were not necessary to fulfil the stated objective of the rule.
- The US had not provided scientific and technical information to demonstrate that the trade-restrictive elements were necessary to fulfil the legitimate objective of the rule.
- The US countered by stating that the TBT was not relevant as the rule was not on the product itself but the company producing it and that the requirement was based on an averaging calculation not on an individual shipment.

4.11.3. *Use of data in dispute settlement*

As one of the later cases, and the first to fall within the jurisdiction of the WTO, it is interesting to note that more data and factual information of a trade and economic nature are provided by both parties in this case but, as with previous cases, it was not considered as part of the ruling by the Panel. Despite the extensive discussions around the validity of Article XX, little environmental information was tabled.

With respect to trade data, the US contended that Venezuela's claims that its gasoline industry was adversely affected by the gasoline rule was not valid as current import volumes had not decreased from historical levels, and also that import volume and domestic production fluctuated depending on market conditions, irrespective of US regulatory action. Further, Venezuela's share of the import market had grown from 11.15% in 1994 to 18.5% in 1995.

The plaintiffs stated that, from a technical point of view, foreign refineries would need to make expensive investments and changes to their refineries to produce a US-conforming product, thus increasing costs of production, or to merge imports with another supplier (selling gasoline at a lower price to them) in order to meet the averaging requirements of the statutory baseline. The US countered the first of these by noting that it could meet the requirements by adding additives such as oxygenates (a low-cost solution), a claim the plaintiffs disagreed with. The US further noted the turbulent economic conditions in the Brazilian refining industry over the period with strikes and recurring labour problems affecting production. They further noted the impracticality from a cost point of view of applying the statutory baseline to domestic producers because of the magnitude of the monitoring requirement to ensure compliance.

4.11.4. *Outcome*

The Panel ruled against the US in this case, for Article III (as imported and domestic products were 'like' products but treated in a discriminatory way) as well as Article XX(b), (d) and (g), and also determined that it did not need to consider the case under the TBT.

With regard to Article XX(b) and (g), the Panel determined the following:

- The gasoline rule was not necessary as there were other means by which environmental measures could be put in to place which were less inconsistent with the GATT/WTO. As part of this, it (i) was not convinced that individual baselines could not be applied to foreign manufacturers, (ii) did not consider the prospect of 'gaming' to be a sufficient reason for differential treatment, and (iii) did not necessarily lead to reduced emissions as there was no volume cap on manufacturers.
- A policy to reduce the depletion of clean air was a policy to conserve a natural resource within the meaning of Article XX(g). However, the Panel saw no direct connection between less favourable treatment of imported gasoline that was chemically identical to domestic gasoline and the US objective of improving air quality and therefore did not consider the measure was primarily aimed at the conservation of natural resources.

4.12. US vs. India, Pakistan, Malaysia and Thailand, Shrimp and Shrimp Products 1998 (WT/DS58 (and 61))

4.12.1. *Introduction*

In 1997, the four countries (above) ('the plaintiffs') brought a joint complaint to the WTO against a ban imposed by the US on the import of certain shrimp and shrimp products. The rationale presented by the US for the ban was that the process used to harvest the shrimp adversely impacted on sea turtle populations (a bycatch) and that they would only accept imported shrimp where turtle excluder devices (TEDs) were used by fishing boats.

This case has many parallels with the Tuna–Dolphin case in that it is an attempt by the US to justify an import ban on the basis of a production/harvest process that does not comply with its domestic laws and norms regarding environmental conservation. Sea turtles are recognised as at high risk of extinction under CITES and signatory countries are duty-bound to avoid imports of products that adversely affect sea turtle populations.

The plaintiff's complaint focussed on lack of compliance under Article XI (quantitative and non-quantitative restrictions on imports) and Article XX(b) and (g). The US declared that its actions were complaint with the General Exceptions article (Article XX). This is the first environmental case to have been dealt with under WTO DSU protocol following the dispute settlement processes outlined in Chapter 2.

4.12.2. *Use of Article XX and other relevant GATT Articles*

In this case, the US cited the preamble to the WTO agreement in its legal argument noting their desire to enable trade but in a manner compatible with the objectives of sustainable development. They argued that the precedent that measures (and rulings under Article XX) that were not able to be applied extraterritorially (as determined in Tuna–Dolphin I) now needed to be reversed in the context of the WTO preamble, and there was considerable discussion in the Panel report on extraterritoriality. The plaintiffs disagreed on this matter, stating that the original sense of Article XX should not be changed by the WTO preamble context, but also noting that the preamble states also that sustainable development should be aspired to "in a manner consistent with the respective needs and concerns... at different levels of economic development".

With regard to Article XX, the plaintiffs stated that it was not appropriate to apply it as, from its preamble measures must not be applied in a manner which would constitute an arbitrary discrimination or a disguised restriction on international trade. They stated that the actions of the US were designed to protect domestic fishing industry from a competitive disadvantage and that the measure was discriminatory as it was applied suddenly to their nations whereas other jurisdictions had been given many years to comply.

The US contended that their measures here were consistent with Article XX(b) 'necessary to protect...' and XX(g) 'relating to the conservation of...'. They noted that the case for conservation is compelling in this case and the measures are 'necessary' and that sea turtles fit the scope of Article XX as they are an exhaustible natural resource. There was discussion here of whether measures to limit the import of one species was relevant in ensuring the conservation of another by the plaintiffs and further discussion on the role of international laws (such as CITES) and the notion of extraterritoriality as well as considerable discussion on the original intent of Article XX and its sub-clauses. The plaintiffs noted that the measures used were not necessary to fulfil US policy and conservation objectives as there were alternatives such as multilateral or bilateral agreements.

4.12.3. *Use of data in dispute settlement*

Substantial data were provided in this case on sea turtle conservation, both the conservation biology and relevant national and international laws to protect them. Each of the plaintiffs provided information on their sea turtle conservation programmes, fishing restrictions and species research programmes which seek to protect the species. The US focussed on data showing large scale species reductions in the plaintiffs' territories over the previous thirty years and demonstrated how most nations prevented incidental turtle catch through the use of TEDs unless manual harvesting was used.

The US further made statements about turtles being a global shared resource, citing migration pattern data and provided data on the role of shrimp trawl fishing in turtle species depletion.

Interestingly, and for the first time, expert views were sought from conservation scientists regarding sea turtles and their ecosystems and this is presented in the Panel report along with many academic references pertaining to turtle conservation.

Economic data were also provided to the Panel in terms of the cost to purchase and fit a TED. The plaintiffs disagreed with the US cost estimates (and them being practical and inexpensive) citing relative income differences between fisherman in their countries and those in the US (with the contention that buying and fitting a TED cost the equivalent of a year's income of $70–600). They further noted the risk of lower catch value as the TED prevented larger fish escaping, and these offered high prices at market that the fisherman would lose out on.

The plaintiffs provided data not only on the size and structure of their fishing fleets (being dispersed and many small-scale fisherman) but also on trade impact arising from the ban. For example, India noted that their exports of shrimp to the US had reduced by about 60% between October and November 1996 because of the ban. They noted the downstream impact of this on industry within coastal communities in India as well as the socio-economic implications in the communities. They further noted the negative impact of increased regulation on cost and speed of export of their products. Perversely, the US countered the data on exports from the plaintiffs with data showing increasing values of shrimp imported from them over the period just before this. The data are obfuscated by differential dates and how values are determined for the traded products.

In this case, the extensive data provided by both the plaintiff and defendant as well as the expert statements were not used in the ruling.

4.12.4. *Outcome*

This case went to appeal, but this upheld the original Panel findings that the US measures were contrary to Article XI and could not be justified under Article XX. It was determined at appeal that the use of Article XX(g) was valid if considered in its own right but was not so when considered in the context of the article's chapeau (which defines when the general exceptions can be cited). In this instance, the US lost the case not because measures to protect sea turtles were illegitimate but because it had discriminated between WTO members. It provided countries in the western hemisphere (mainly the Caribbean) with technical and financial assistance and longer transition periods for fishermen to start using TEDs.

4.13. Canada–EU, Asbestos and Asbestos-Containing Products 2001 (DS135)

4.13.1. *Introduction*

In this dispute, Canada challenged a decision by France to place a ban on import and use of chrysotile asbestos and products containing it. The rationale for this was that as a highly toxic material used in various industrial sectors, it posed significant health threats to human health. It noted its health risk not only to construction workers subject to prolonged exposure but also to the general population subject to occasional exposure.

Canada argued that, first, there should be a distinction between chrysotile asbestos itself and products containing it (such in cement) where the product contained no health risk, and also argued that France's actions violated GATT Articles III:4 and XI as well as aspects of the TBT (within Article 2) and nullified and impaired benefits under GATT XXIII:1(b). The EU (representing France) countered these by stating that this was not covered by the TBT and that if not compatible elsewhere under GATT should be considered justified with Article XX(b).

4.13.2. *Use of Article XX and other relevant GATT Articles*

Article XX was considered specifically in terms of sub-clause (b) 'necessary to protect human health' and also the chapeau requiring the measure to be least trade-restrictive and non-discriminatory.

The Panel determined that there was sufficient evidence that chrysotile asbestos was substantially harmful to human health to merit its ban under clause (b), whereas the use of substitute products had not been shown to pose the same risk.

In terms of the chapeau, the Panel explored alternatives such as policies and regulations regarding its use but concluded that such measures could not afford the same protection to health. For example, while regulations could be put in place for the construction industry, the same could not be applied to the domestic Do It Yourself (DIY) enthusiast. Also, the measures were a ban on the substance rather than a restriction on the importation, as such it was not a disguised restriction.

4.13.3. *Use of data in dispute settlement*

Trade and economic data are noted as part of the information provided to the Panel. It is noted that global production of chrysotile asbestos was 2 million tonnes in 1997 and Canada was the second largest producer (after the Commonwealth of Independent States). Canada is the largest exporter worldwide at 4,30,000 tonnes of which, prior to the ban, 20,000–40,000 went to France. The industry is valued at 225 million Canadian dollars and provides around 4,000 jobs in Canada, largely Quebec. Relatively little of the substance is used within country, such that the EU noted that Canada 'is exporting to other countries a public health risk'.

There was substantial data provided on the relative risk to human life and health of asbestos. For example, France noted that the number of serious illness (morbidity) had increased four-fold between 1985 and 1995, costing the government 1.8 billion (French) Francs in 1997 (40% of the total occupational health compensation budget). Canada recognised the long-time lag between exposure and effect, suggesting such data related to exposure years/decades previously and provided counter data showing relative risks per 10,000 inhabitants, showing the current risks associated with working with asbestos (with current management and regulatory practices in place) are less than 1, compared with higher risks of moderate alcohol consumption, risks from hurricanes, drinking tap water, etc. Further, substantial data were provided from expert bodies, lawyers and scientists on the relative risks to human life and health from exposure to chrysotile asbestos.

The Panel in this case did recognise the relevance of the scientific data relating to human health and took this into account. No mention was given on trade or economic implications of their decision.

4.13.4. *Outcome*

The Panel ruled in favour of France/EU despite its violation of Article III was it was found to be justified under Article XX(b), considering it necessary to protect human life and health. Under appeal, this finding was substantiated despite the Panel determining that the dispute should have been considered under the TBT rather than the GATT rules. For this, the Panel looked to precedent (e.g. from the Gasoline and Shrimp–Turtle cases) on the use of 'human health risk' and 'necessary' to determine compliance with the clause.

4.14. EU–Brazil, Retreaded Tyres 2007 (DS332)

4.14.1. *Introduction*

The EU brought a case against Brazil to the WTO regarding discriminatory behaviour in relation to retreaded tyres exported from the EU. Brazil had put in place an import ban, fines for marketing transportation, storage, keeping or warehousing of imported retreaded tyres (but not domestic ones), but provided exemptions to imports and fines of retreaded tyres from MERCOSUR countries (Argentina, Paraguay, Uruguay, and Venezuela). At the time Brazil imported both retreaded tyres from the EU as well as used tyres to be retreaded in Brazil before sale — it was only the former that were subject to the import ban.

The EU contended that these actions were inconsistent with WTO obligations in relation to Articles I, III, XI and XIII of GATT (least favoured nation, discrimination, quantitative restrictions, national treatment). Brazil countered these and also claimed their actions were justified under Article XX(b) (general exceptions necessary to protect human life and health).

4.14.2. *Use of Article XX and other relevant GATT Articles*

Brazil contended that its actions were covered by Article XX(b) being necessary to protect human life and health. Precedent was used (US-gasoline and EU-Asbestos, both above) for the justification of reused tyres being within scope. Brazil's argument for it being within scope centred on issues of waste accumulation and disease. It noted that retreaded tyres have a shorter lifespan than new tyres but were attractive in a price sensitive market such as Brazil. This meant there was a likelihood of higher levels of tyre waste accumulation. It was further noted that waste tyres brought two types of environmental risk: first, they increased breeding grounds for mosquitoes (creating pools of still water), increasing epidemics of mosquito-borne diseases (malaria, dengue fever, yellow fever, etc.) but also release toxic chemical and heavy metals, with increased health risks from contamination, including cancer. The EU disputed the direct link between retreaded tyres and human life and health, noting it was the mechanism of disposal, not the product itself that created the risk. As such, it disputed both the concept of necessity of the measure as well as the direct link between retreaded tyres and human life and health, and considered that there were other, less trade-restrictive means by which to achieve the objectives.

4.14.3. *Use of data in dispute settlement*

Substantial amounts of information and data were provided in this case both from the plaintiff and defendant as well as third party nations and relevant experts. The majority of the information and data related to the technicalities of retreading tyres, in terms of process, quality of the product, potential retreading rates and waste implications. In addition, there was considerable discussion on the potential to retread tyres based on their quality and provenance.

Statistics were provided by Brazil relating to the epidemiology of various malaria-borne diseases in the country (numbers reporting each disease over time). In addition, Brazil cited research evidencing the attractiveness of tyres as breeding grounds for mosquitoes, and their ability to withstand transportation, thus risking spread of disease throughout the country. However, they could not provide substantiated evidence of the numbers of waste tyres in the country (they 'assumed' it was about 100 million).

There was discussion, but no data on economic and trade implications. Data were provided on the numbers of tyres imported to Brazil by the EU. Brazil noted that, as a less wealthy country, its price sensitivity for tyres was higher than in the EU and that the EU was using its waste surplus in used tyres as an exportable product to Brazil, rather than bear the costs of disposal in country. Brazil noted that, as retreaded tyres had a shorter life span, they were, effectively, bearing the pollution costs for the EU of used tyres.

4.14.4. *Outcome*

The Panel concluded that the measures taken by Brazil were inconsistent with a range of GATT articles, including Article XI (prohibition on quantitative restrictions) and Article III (national treatment). With regard to Article XX, the Panel concluded that the measures were 'necessary' within the meaning of Article XX(b) and that the subject of the dispute was legitimately within 'human health and life'. However, it did not consider that the approach used by Brazil was the least trade-restrictive alternative and that there were other reasonably available alternatives to an import ban. For this, it will have used information provided on the retreading industry processes as contextualisation. In addition, through the appeals process, it was determined that the Mercosur arrangement (whereby the import ban and other measures did not apply in the same manner as the EU) constituted arbitrary or unjustifiable discrimination and a disguised restriction on trade within the meaning of the chapeau to Article XX.

4.15. Summary of Analysis

Each of the 12 ecosystem services-related cases reviewed have applied the General Exceptions Article (Article XX(b) and/or (g)) within the dispute settlement process, with eight of the cases coming before GATT and the balance under the aegis of the WTO. Throughout the period, some precedents and common understandings have occurred, including the following:

- a definition of 'necessary' in Article XX(b) and 'relating to...' in Article XX(g);
- a common understanding of the term 'exhaustible natural resource' and attributes considered within its scope;

- an agreement as to the geographical scope of influence of any trade dispute and whether one nation can impose its environmental standards on another's territory.

As a general trend, it is clear that the use of data (economic, social[1] and/or environmental) has been considered relatively insignificant in the Panels' decision-making, although the presentation of data has increased in the Panel proceedings over time. A summary of extent of data presented and its consideration in decision-making is provided in Table 4.1. There are, however, very few instances where data have been considered significant to the Panel deliberations and, where it has been used, it has been inadequate as a sole basis for a decision (e.g. in Gas Guzzler/CAFE and Reformulated Gasoline trade impact was mentioned in the Panel deliberations). Where data have been provided, they are sporadic in nature and in no case provide a holistic picture of the economic–social–environment effects of the trade dispute in question.

The scoring system used in the table is as follows:

0 = No data provided.
1 = Some data provided but not extensive; not considered in Panel decision-making.

Table 4.1. Summary of adequacy and relevance of data for ecosystem services cases.

Case	Economic	Environmental	Social
Canadian Tuna 1982	0	1	0
US Superfund 1988	2	1	0
Canada Herring and Salmon 1988	1	1	0
Thai Cigarette Imports 1990	2	0	1
Mexican Tuna–Dolphin I 1991	0	2	0
Canadian Beer 1992	1	1	0
EU Tuna–Dolphin II 1994	0	2	0
US Gas Guzzler and CAFE 1994	3	2	0
Venezuelan Reformulated Gasoline 1996	3	1	0
India, Pakistan — Shrimp–Turtles 1998	2	2	1
Canada Asbestos and Asbestos-Containing Products 2001	1	0	3
Brazil Retreaded Tyres 2007	1	2	2

[1]In the context of these cases, 'social' refers to the impact on human health.

2 = Extensive coverage of relevant data; not considered in Panel decision-making.
3 = Some data provided but not extensive; used in Panel decision-making.
4 = Extensive coverage of relevant data; used in Panel decision-making.

4.16. Limitations of the WTO's Approach to Date and Alternative Options

There are a number of limitations to the rules-based approach used by the WTO to date in resolving ecosystem services-related trade disputes. The core problem is the potential inability for a rules-based approach to deliver against organisational objectives. The WTO's objectives are to maximise social[2] welfare through the mechanism of least trade restrictiveness, but the rules as they stand, and their interpretation through the dispute resolution process, has focussed on least trade restrictiveness *per se*. As seen in the cases above, little regard is given to the economic, environmental and social impacts of a ruling. Within this, other problems emerge. These include the following:

- The WTO's stance on extraterritoriality (see Dolphin–Tuna cases) where one nation cannot restrict imports from a nation regardless of the impact of its production methods on the environment (and their inherent subsidy to costs).
- The rules as they stand consider the 'intent' of an ecosystem services action and not its impact or 'effect' (Esty, 1994).

Over time, there have been various suggestions as to how the current 'rule-based' decision-making system could be altered to ameliorate current evaluation problems. These have broadly involved either suggestions as to how the rules could be altered or supplemented, or changes to the Panel processes or removal of ecosystem services-related cases to another organisation (Esty, 1994; Cameron and Ward, 1994; Charnovitz, 1993).

The first of these, changing the rules, has been the most popular option. There have been calls for a proportionality clause so that any proposed measure could only be put in place if it was a reasonable cost to do so (Trebilcock and Howse, 1995), revised interpretation of 'product' to include

[2]Social welfare is used to encompass human wellbeing not just of economic impacts but also environmental and social as consistent with Article XX and definitions of sustainable development.

its production method (Thaggert, 1994), inclusion of an Environmental Impact Assessment into the proceedings (Cameron and Ward, 1994). Esty (1994) wrote extensively on the need to change the Panel process and called for a 'three-pronged' approach considering each of: (i) intent and effect, (ii) environmental legitimacy, and (iii) unjustified trade disruption through a major revision of the 'rules' rather than through a call for a data-evaluation approach.

Despite these calls, the WTO has continued to approach trade–ecosystem services disputes through interpretation of Article XX with scant attention given to economic, environmental or social impacts of trade. We contend here that in order to understand the holistic dispute under consideration, comprehensive data analysis of such impacts is required (Haddock, 2000). There are many advantages to such an approach. First, it widens the scope of what is considered relevant to the decision and as such increases the chance that the Panel will determine an outcome in line with WTO objectives. Second, it enables quantification and qualification of the facts to be considered and therefore the ability to assess relatively between impacts. Previously, there have been practical barriers to implementing this approach, not least the ability to obtain meaningful data, to assess it based on rigorous valuation techniques and in determining acceptable levels of uncertainty (Haddock, 2000). Two decades on from such limitations being identified, Part C of this book demonstrates how a data-led approach to analysing ecosystem services-related trade disputes is now both possible and beneficial to social welfare.

Part B

Evaluating International Trade Proposals in the Context of Biosecurity Risks

Chapter 5

Evaluating International Trade Proposals in the Context of Biosecurity Risks: Methodology

5.1. Introduction

As outlined in Chapter 2, Article 5 of the SPS Agreement identifies the factors considered relevant from a WTO perspective in assessing the extent of quarantine risks, and typically involves either quantitative or qualitative assessments of possible invasive species impacts. However, as discussed in Chapter 3, omitted from this list of relevant factors is *consumer gains from trade*. This is problematic for any assessment of the relative benefits and costs of a trade decision as the benefit side of the ledger is effectively missing. Specifically, in considering a proposal to allow imports of a product, analytical models informing a decision should reconcile the net benefits of trade with the potential costs of an outbreak of invasive species associated with trade. As a consequence, it was argued in Chapter 3 that the core problem with the WTO's trade dispute resolution process is the potential inability of the WTO's rules-based approach to deliver against its organisational objective of maximising social welfare.

Previous literature involving the economic modelling of trade and invasive species invasions is a bit of a mixed bag. For example, modelling efforts have been devoted to the prevention of invasive species introductions from international trade (Costello and McAusland, 2003; Horan *et al.*, 2002a) or to the control of already established invasive species (Olson and Roy, 2002; Eiswerth and Johnson, 2002). More recently, integrative approaches where the trade-offs between prevention and control of invasive species are considered have been developed (Finnoff *et al.*, 2007; Kim *et al.*, 2006). However, these studies rarely include a welfare analysis where the benefits of trade to consumers are considered.

As a consequence of these limitations, this chapter develops a conceptual framework that takes into account both potential gains from trade and changes in invasive species risk when making trade decisions. In trade dispute situations, economic analyses designed around this framework can be used to determine if the net benefits of imports are likely to exceed the quantifiable increase in pest damage, and therefore if trade will result in a net gain for society. If not, refusing a market access request is justifiable on the basis of preventing a net social welfare loss.[1]

The structure of this chapter is as follows: Section 5.2 develops the methodological framework for incorporating the biosecurity risk of an invasive species introduction, as well as the impacts on consumer and producer surplus, into the total welfare evaluation of the impacts of trade; Section 5.3 analyses the range of possible outcomes from this evaluation which takes account of both the gains from trade impacts and the invasive species impacts; and Section 5.4 concludes the chapter by summarising its contents and leading into the two case study applications of this methodology presented in Chapter 6.

5.2. Theoretical Model for Welfare Analysis

Consider the case of the WTO deciding whether or not to permit the import a homogenous good into one country from another country.[2] This good has the potential to act as a pathway for a harmful host-specific pest or disease that the source country has but the importing country does not. Assume that in the absence of price-inflating Sanitary and Phytosanitary Standard (SPS) measures, the landed price of imported product (p^{**}) is below that of a domestic equivalent (p_0), and that the domestic market is small relative to the rest of the world in terms of its influence on the world price. The domestic market for the product is characterised by a downward

[1] The framework presented in this chapter is based on that developed in Cook and Fraser (2008).

[2] In the following section, no explicit allowance is made for the net gains from trade for the exporting country in this decision-making process. However, it is straightforward to include these as a consideration — as is done in Section 5.3 and for the case studies in Chapter 6. See also Cook and Fraser (2008).

sloping demand curve, (q) , and an upward sloping supply curve, $g(q)$. This situation is depicted in Figure 5.1, the details of which are explained in the following.

The domestic losses that could result from an exotic pest or disease outbreak resulting from contaminated imports can be estimated as the total expected change in producer surplus brought about by an incursion-induced (negative) supply shock, plus the cost of controlling the species (be it eradication or suppression). The probability of arrival (γ) is most likely an increasing function of the quantity of imported product, q^*, and a decreasing function of the pre-border and border SPS measures the good is subjected to in the process of importation with cost c_{I} (i.e. $\gamma(q^*, c_{\mathrm{I}})$). To simplify the effects of uncertainty, it is useful to assume a deterministic change in the probability of arrival with SPS-compliant imports from abroad (γ^*) relative to the probability of arrival without imports (r) (i.e. $\gamma^* > \gamma$).

As a starting point, a closed economy involves domestic producers with a supply schedule $g(q)$ providing the total supply (q_0) to the domestic market at a price p_0. If an incursion were to occur (despite there being no trade pathway) the supply curve will shift inwards to $h(q)$, and the new equilibrium price will rise to p_1 at which q_1 will be demanded. Note that even when no trade takes place $0 < \gamma < 1$. Note also that for clarity of exposition we use a static model and thus describe the shifts of the supply curve as an instantaneous process.

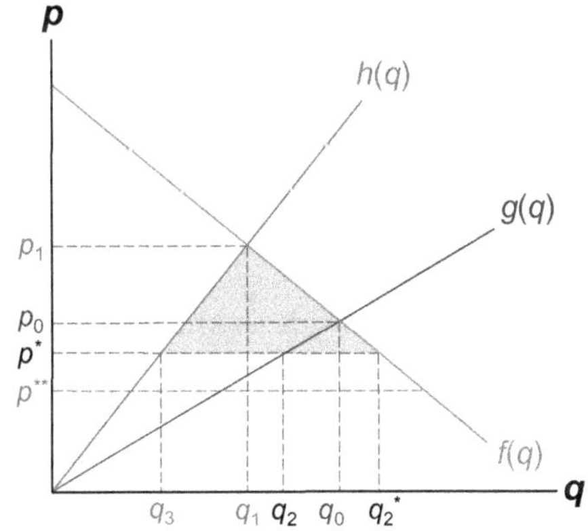

Figure 5.1. The quarantine-restricted trade decision from a closed-economy position.

In reality, the shift of the supply curve is dynamic and linked to the spatial spread of the invasive organism. Note that the bioeconomic models described in Chapter 6 account for the dynamic interaction between spread and the modifications of the domestic supply curve.

If the market were to move from a closed situation to a quarantine-restricted trade situation, the prevailing market price will fall to the world price (p^{**}) plus c_{I} (i.e. $p^* = p^{**} + c_{\mathrm{I}}$ where c_{I} is sufficiently low to ensure that $p_0 > p^*$). Domestic producers will remain suppliers to the domestic market as long as p^* remains above the minimum average variable cost of production, supplying a lower quantity q_2. However, if trade takes place the probability of contaminated product reaching the importing country via the trade pathway provided by $q_2^* - q_2$ imports increases from γ to γ^*. In the event that a pest or disease incursion does result the supply curve will shift inwards to $h(q)$, further reducing the quantity supplied by domestic producers to q_3.

Following an incursion, a coordinated control campaign is mounted against the invasive species to either eradicate it or restrict its abundance and distribution. Assume that the total cost of control will depend only on the size of the outbreak upon detection (s) and the total reduction in abundance and distribution sought by the campaign (a), and is denoted $c(s, a)$ (Olson and Roy, 2005). Total control costs are assumed to be increasing in both a and s, while marginal control costs are increasing in a and non-increasing in s.

On this basis, *Expected Impact* of the invasive species if no trade takes place (EI_{A}) is given by

$$EI_{\mathrm{A}} = \gamma \left[\left(p_0 - \int_0^{q_0} g(q) \right) \mathrm{d}q - \left(p_1 - \int_0^{q_1} h(q) \right) \mathrm{d}q + c(s,a) \right] \quad (5.1)$$

Equation (5.1) states that EI_{A} is equal to the expected difference between the producer surplus under autarchy if no outbreak occurs and the producer surplus under autarchy if an outbreak occurs, plus the expected cost of control.

Secondly, the decision-maker needs to know the expected impact of the invasive species if trade takes place (EI_{Q}):

$$EI_{\mathrm{Q}} = \gamma^* \left[\left(p^* - \int_0^{q_2} g(q) \right) \mathrm{d}q - \left(p^* - \int_0^{q_3} h(q) \right) \mathrm{d}q + c(s,a) \right] \quad (5.2)$$

Equation (5.2) states that EI_{Q} is the expected difference between the producer surplus with trade if SPS measures are 100% effective and the producer surplus if an outbreak occurs, plus the expected cost of control. Note that the prevailing market price in the event of a disease incursion is assumed to be p^* (rather than p^{**}) due to the non-specificity of SPS measures. The imported commodity will in general entail a risk of introduction of multiple pests and, even if one of them becomes naturalised, we assume that c_{I} still applies.

It follows that the expected cost to producers of the increase in invasive species risk brought about by trade (EI^*) is given by

$$EI^* = EI_{\mathrm{Q}} - EI_{\mathrm{A}} \tag{5.3}$$

Despite the risk of pests and disease hitchhikers, the importation of goods from abroad also brings with it potential gains from trade by placing downward pressure on prices. Imports provide a greater quantity of the good to consumers at a lower price, p^*, compared to the closed economy (autarchy) situation which, as shown in Figure 5.1, imposes costs to consumers and provides gains to domestic producers. Therefore, the decision of whether to import a commodity subject to SPS measures must be made relative to a closed economy situation and must also establish the consumer benefits achieved and producer costs imposed by permitting trade.

In moving from a closed economy to a quarantine-restricted trade situation, the prevailing market price would be expected to fall to p^* at which domestic producers are willing to supply q_2 of the total quantity demanded, q_2^*. The total consumer surplus gained by allowing quarantine-restricted trade is given by

$$\Delta CS = \left(\int_0^{q_2^*} f(q) - p^*\right) \mathrm{d}q - \left(\int_0^{q_0} f(q) - p_0\right) \mathrm{d}q \tag{5.4}$$

This change in consumer surplus is the difference between the post-quarantine trade consumer surplus and autarkic consumer surplus. This gain comes at the cost of competition-induced producer surplus losses to domestic producers:

$$\Delta PS = \left(p_0 - \int_0^{q_0} g(q)\right) \mathrm{d}q - \left(p_0 - \int_0^{q_2} g(q)\right) \mathrm{d}q \tag{5.5}$$

This producer surplus change is calculated as the difference between the autarkic producer surplus and post-trade producer surplus. The resultant net gains, termed *traditional gains from trade* in Snape and Orden (2001), are simply the difference between consumer surplus gain and producer surplus loss, ignoring the possibility of an invasive species incursion. That is, the traditional gains from trade here represent the change in producer and consumer surplus as a result of price differentials between the domestic equilibrium price and landed price of imports:

$$GT = \Delta CS - \Delta PS \tag{5.6}$$

This is represented as the shaded region in Figure 5.1.

5.3. Using the Model for Welfare Analysis

As discussed earlier, a complete welfare evaluation of a trade proposal should take into account not only these traditional trade effects but also the effects on producers of the increase in invasive species risk (EI^*) brought about by trade (i.e. $EI^* = EI_{\mathrm{Q}} - EI_{\mathrm{A}}$). Therefore, the total gains to consumers resulting from trade must be sufficiently high to offset all the expected losses to domestic producers for there to be a net gain from moving from a closed economy to a quarantine-restricted trade setting.

More specifically, combining the changes in consumer and producer surplus with the expected impact of an invasion on producers, the expected net gains from trade (NG_{E}) can be stated as

$$NG_{\mathrm{E}} = GT - EI^* \tag{5.7}$$

It follows that the decision of whether or not to import the product based on both the gains from trade and the associated biosecurity risk is either:

(a) if $GT - EI^* > 0$, then it is welfare improving to allow trade to occur;

or

(b) if $GT - EI^* < 0$, then it is welfare improving not to allow trade to occur.

However, as noted in Section 5.2, the gains from trade in such an evaluation consider only the market impacts for the traded good in the importing country. Using an analogous diagram to that represented in Figure 5.1, but instead designed to represent the exporting country, the market clearing price is presumably below the price at which it supplies to the importing country. Therefore, when trade takes place, the domestic price of the exported good rises, and with domestic demand now supplemented by exports to consumers in the importing country, the total quantity of the good sold by producers is greater.

As a consequence, consumer surplus in the exporting country (CS_X) is set to decline due to the price rise induced by redirecting a proportion of supply to the importing market, but this is more than offset by the rise in producer surplus in the exporting country (PS_X). The traditional gains from trade for the exporting country (GT_X) can therefore be stated as

$$GT_X = \Delta PS_X + \Delta CS_X \tag{5.8}$$

If these exporting country gains from trade were to be added to the importing country gains from trade, then total gains from trade (i.e. $GT_T = GT + GT_X$) can instead be used in determining the expected net gains from trade (NG_E), which can be restated as

$$NG_E = GT_T - EI^* \tag{5.9}$$

In this case, it follows that the social welfare of the global economy could be improved by allowing trade even where the combined net impacts of the gains from trade and the increased biosecurity risk in the importing country are negative. This potential situation is considered further in the context of the case studies presented in Chapter 6.

5.4. Conclusion

This chapter has been motivated by the criticism in Chapter 3 that the core problem of the WTO's rules-based approach to trade dispute resolution is the potential inability for this approach to deliver against its organisational objective of maximising social welfare. In particular, this rules-based approach as it stands does not adequately evaluate the main impacts of trade on producers and consumers, or the impact of the increased biosecurity risks associated with trade.

In an attempt to take account of these limitations, this chapter has developed a conceptual framework on the basis of which trade dispute decisions that account both of the potential gains from trade and of the associated impact on biosecurity risk can be made, and in so doing enable decisions which improve social welfare.

In Chapter 6, we apply this methodology to two of the biosecurity-related trade dispute cases identified and discussed in Chapter 3 in order to illustrate how the methodology can be used to improve the basis on which trade dispute decisions are made.

Chapter 6

Evaluating International Trade Proposals in the Context of Biosecurity Risks: Case Studies

6.1. Introduction

In this chapter, we select some of case studies presented in Chapter 3 and illustrate the application of the methodology developed in Chapter 5 to evaluate the impacts of allowing imports on domestic producers and consumer surplus and on biosecurity risks to domestic production. The first case study involves New Zealand (NZ) apples imported into Australia, and the second fresh salmon imported into Australia from Canada. Recall that in both cases, the import of the goods represented an increased risk of disease outbreaks within the competing domestic industries. Following the methodology set out in Chapter 5, we use the traditional gains from trade as a benchmark with which to compare the risk of disease in domestic production losses resulting from trade to indicate if trade is likely to be welfare enhancing or retarding. However, in order to allow for the inherently dynamic biological processes involved in disease outbreaks, in what follows we will use Monte Carlo simulations of a bioeconomic model to examine a range of possible scenarios and evaluate the likely net welfare effects of allowing trade.

Note that in both case studies we will not be estimating the gains from trade for the exporting country; however, consideration of these gains will be made in the discussion of the results.

6.2. New Zealand Apples Importation Request to Australia and the Risk of Fire Blight Introduction

6.2.1. *Background*

The matter of importing NZ apples into Australia, and the analysis of this issue, has considerable history to it. We covered some of the technical matters of this case in Chapter 3 but provide some more of the background in this section. We begin in 1989 when NZ made a formal market access request to import apples into Australia. Australia's detailed response, coordinated by the Australian Quarantine and Inspection Service (AQIS), was partly motivated by the so-called 'Lindsay Review' of Australian quarantine (DPIE, 1988) that recommended moving away from 'zero risk' policies towards 'acceptable' quarantine risk. The economic component provided by Hinchy and Low (1990) was accompanied by a biological component (Roberts, 1991). The former took the form of a benefit–cost analysis comparing the expected consumer and producer surplus changes resulting from relaxing quarantine laws protecting the apple industry. The major disease transference concern was considered to be fire blight.

Fire blight is a disease caused by the bacteria *Erwinia amylovora* that affects plants from the family *Rosaceae*, including apples and pears. Once established, the bacteria cannot be eliminated from an orchard, but costly measures such as an aggressive pruning regime can be taken to limit the extent of infection (Buckner, 1995). The disease originated in the United States and has spread to most apple-growing regions of the world with the exception of Australia. It was first discovered in NZ in 1919, and a ban on entry to the Australia fresh apple market was put in place in 1921 (Biosecurity Australia, 2004).

In 1995, NZ made another request to access the Australian apple market. This time, the economic analysis, detailed in Bhati and Rees (1996), was quite different in approach to that of Hinchy and Low (1990). Expected consumer surplus change was not discussed. The analysis only considered possible producer surplus losses to pome fruit growers if a fire blight outbreak were to occur. Both import access requests were denied. Viljoen *et al.* (1997) present evidence that the import ban was indeed justified given that the pear industry in Australia could collapse in the event of a fire blight outbreak.

NZ again submitted an application to Australia to access the apple market in 1999 in which it asked specifically for management procedures that might be applied in order to reduce the risk of biological contamination below Australia's Appropriate Level of Protection (ALOP). Following the release of a draft import risk analysis in February 2004, which recommended that market access be granted subject to strict pre-entry quarantine measures, mistakes were identified in the spreadsheet models used to ascertain the risks associated with certain pests. When these mistakes were corrected and the final import risk analysis was released in late 2006, market access was still deemed to present a sub-ALOP level of risk, but only if an even more stringent set of SPS measures were applied.

These restrictions were considered by NZ to be inconsistent with Australia's obligations under the SPS Agreement. In 2007, NZ requested consultations with Australia using the WTO's Dispute Settlement Process (DSP), and this is where we took up the story in Chapter 3. The European Union and the United States requested to join the consultations. Despite the DSP initially targeting a completion date of June 2009, it was not until August 2010 that the Panel handed down its verdict and declared Australia's restrictions to be in breach of the SPS Agreement. Australia unsuccessfully appealed this decision.

Section 6.2.2 applies a partial equilibrium model that combines trade with a stratified dispersal model for the spread and control of fire blight in Australia. This represents a coherent economic framework of analysis to the empirical context of importing NZ apples, and aims to evaluate the social welfare consequences of allowing this trade.

6.2.2. *Modelling the supply curve shift in the event of fire blight introduction*

To estimate the invasion costs in both the autarkic and quarantine-restricted trade scenarios, a biological model is used to simulate the arrival, spread and impact of fire blight in Australian apple orchards over a 30-year time period. This model is then combined with a measure of the marginal damage cost of invasion and periodic eradication attempts.

The biological model generates fire blight entry events using a Poisson process, which is often used to model the entry of invasive species and diseases to new environments (Vose, 2000). The Poisson stochastic process implicitly assumes that each introduction is independent of another. The single parameter used to define the Poisson distribution is the expected number of introductions per time step (i.e. in our case 1 year), λ.

With the exception of well-known, high-profile invasive species, quantitative arrival probability estimates are difficult to come by and often are poorly specified. We use an estimate put forward in Cook (2003) for the likelihood of fire blight arrival, which in turn uses Australian government guidelines for risk analysis (Biosecurity Australia, 2001) to represent λ using a uniform distribution. From Cook (2003), we assume that, on average, one introduction event would occur every 2 years under an SPS-restricted trade scenario. However, this may be as high as one introduction event every year, or as low as one every 3 years.

We therefore define λ under SPS-restricted trade using a Uniform distribution with a minimum value of 0.3 and a maximum value of 0.7, i.e. Uniform $(0.3, 0.7)$. Under autarky, we assume that λ falls to Uniform (0.001, 0.05). Table 6.1 summarises these and other parameter estimates used in the model and explained in the following.

In iterations of the model in which the Poisson process predicts a successful introduction, the probability of that event leading to a successful establishment is assumed as Uniform $(0.7, 1.0)$. This indicates that although introduction does not necessarily lead to establishment, in the case of fire blight there is a high probability that it will.

Indeed, the likelihood of establishment in Australia has recently been shown to be very high using Self-Organising Map (SOM) analysis, which is a type of artificial neural network. This technique uses worldwide species associations to determine which species have the highest likelihood of being established (Worner and Gevrey, 2006). An SOM analysis was performed on the worldwide distribution of 131 bacterial pathogens (CABI/EPPO, 2003), of which 71 are currently absent from Australia. The analysis ranked fire blight 17th in this list (Paini *et al.*, 2010). Thus, we considered a high probability of establishment with a median of 0.85 (Table 6.1).

Once entry and establishment have occurred, spatial homogeneity is assumed in the sense that different host plantings across the country are assumed equally susceptible to infection. If uncontrolled, fire blight will continue to spread to the point where it would become *naturalised.* Naturalisation is complete when a species spreads to its full capacity within an environment, such that descendants of the original specimens introduced into that environment become permanent, non-spreading members of the flora/fauna (Mack, 1996; Mack and Lonsdale, 2001).

The area of infection is generated by a reaction diffusion model derived from Fisher (1937). This form of model has since been expanded for more general use (Britton, 1986; Skellam, 1951), and has been shown to approximate the spread of a diverse range of organisms (e.g. Okubo and Levin, 2002; Dwyer, 1992; Holmes, 1993; McCann *et al.*, 2000).

A generic result is that an infection diffusing from a point source will eventually reach a constant asymptotic radial spread rate of $2\sqrt{rD}$ in all directions, where r is the infection's intrinsic rate of growth. Hence, we assume that once established the infection spreads by a diffusive process such that the area affected expands following the function (Hengeweld, 1989; Lewis, 1997; Shigesada and Kawasaki, 1997)

$$A_t = \pi(2t\sqrt{rD})^2 = 4D\pi rt^2 \tag{6.1}$$

Here, A_t is the area of hosts infected with fire blight at time t, D is the infection diffusion coefficient and r is the intrinsic rate of infection growth.[1] In each unit of area infected, the local infection density $N(t)$ grows over time following a logistic function until the carrying capacity of the environment (K) is reached.

$$N(t) = \frac{K}{1 + \left(\frac{K}{N_{\min}} - 1\right) e^{-rt}} \tag{6.2}$$

Here, $N_{\min}$ is the size of the original infection.

The area and density functions of Equations (6.1) and (6.2) are combined with a logistic function generating changes in the number of nascent foci (or *satellite* sites) as the process of infection continues. These are events that 'jump' the expanding infection beyond the invasion front and are attributable to causes external to the invasion itself such as weather events or animal or human behaviour. The number of new satellites created in each unit of time ($s(t)$) is

$$s(t) = \frac{s_{\max}}{1 + \left(\frac{s_{\max}}{s_{\min}} - 1\right) e^{-\mu t}} \tag{6.3}$$

Here, μ is the intrinsic rate of new foci generation, $s_{\max}$ is the maximum number of satellite sites generated in a single time step (i.e. 1 year) and $s_{\min}$ is the minimum number of satellite sites generated per time step (Moody and Mack, 1988). Once a satellite site is established, the infection begins to expand in the same manner as the original site. The total combined area of

[1] For analyses performed at higher spatial resolutions, a relatively simple modification can be made to constrain the rate of spread according to area and shape of potentially suitable habitat, or the distribution of the host. In our simple case, Equation (6.1) allows the prediction of spread on the basis of a constant rate of infection. An estimate of D can be derived from the mean dispersal distance ($\bar{d}$) where $D = 2(\bar{d})^2$ (Andow *et al.*, 1990).

infection grows until $A_t = A_{\max}$ (maximum habitable area), at which point total area remains constant. This point represents the *carrying capacity* of the environment. Environmental and demographic stochasticity leading to non-uniform invasion is not considered in the model.

The complete model simulates the ecological processes of arrival, establishment, spread and impact using the values and statistical distributions given in Table 6.1.[2]

In terms of preventing naturalisation, eradication is the only government management activity simulated in the model. It involves the complete removal of infested trees and the creation of intensive buffer zones (using fungicide treatments) around infested sites (WAQIS, 1999). In Australia, fire blight is a listed species under the Emergency Plant Pest Response Deed (EPPRD) (PHA, 2005), which states that in the event of an incursion a pre-agreed cost-sharing arrangement for eradication is activated.[3] This aids in the early reporting of an incursion, minimises the response time and capitalises on the opportunity to minimise the size and impact of an incursion. Eradication is the only management activity covered under the EPPRD and is conditional on results of a technical feasibility study and a benefit–cost analysis (PHA, 2001). If eradication is deemed infeasible or likely to result in a net loss, the Deed no longer applies and much of the management responsibility falls to affected industries. However, once an eradication campaign has begun, it is largely technical and logistical factors that determine if it will continue in the long term.

We assume that eradication is immediately commenced once the industry and government have been alerted to the presence of fire blight in Australia. The detection that triggers the EPPRD is, on average, assumed to occur in 60% of incursion events simulated by the model using a binomial distribution (i.e. binomial (1.0, 0.6)). The probability that the eradication

[2]Due the uncertainty surrounding some of these parameters, a range of distributional forms are specified in Table 6.1. They include (a) pert — a type of beta distribution specified using minimum, most likely (or skewness) and maximum values; (b) uniform — a rectangular distribution bounded by minimum and maximum values; (c) binomial — returning a zero (failure) or one (success) based on the number of trials and the probability of success.

[3]Listed species fall into one of four cost-sharing categories relating to their potential impacts on public and private resources. The category chosen dictates an appropriate split of eradication funding between government and private funding sources. Currently, fire blight is classified as a category two species, indicating a 20% private and 80% government/public funding contribution (PHA, 2005).

Table 6.1. Parameters and their assumed values.

Parameters	Quarantine Restricted Trade	Autarchic Trade
Biological		
Detection probability	Binomial (1.0, 0.6)	Binomial (1.0, 0.6)
Infection diffusion coefficient, D (ha/yr)[c]	Pert (1.0, 1.5, 2.0)	Pert (1.0, 1.5, 2.0)
Intrinsic rate of infection and density growth, r[e]	Pert (1.0, 1.25, 1.5)	Pert (1.0, 1.25, 1.5)
Intrinsic rate of new foci generation per unit area of infection, μ[c]	Pert (0.01, 0.03, 0.05)	Pert (0.01, 0.03, 0.05)
Maximum area infected, $A_{\max}$ (ha)[d]	32,760	32,760
Maximum infection density, K (#/ha)[c]	Pert ($1.0 \times 10^6, 5.5 \times 10^6, 1.0 \times 10^7$)	Pert ($1.0 \times 10^6, 5.5 \times 10^6, 1.0 \times 10^7$)
Maximum number of satellite sites generated in a single time step, $S^c_{\max}$	Pert (50, 60, 70)	Pert (50, 60, 70)
Minimum area infected immediately upon entry, $A_{\min}$ (ha)	1	1
Minimum infection density, $N_{\min}$ (#/ha)	1	1
Minimum number of satellite sites generated in a single time step, $S_{\min}$	1	1
Probability of entry[a]	Uniform (0.3, 0.7)	Uniform (0.05, 0.3)
Probability of establishment[b]	Uniform (0.7, 1.0)	Uniform (0.7, 1.0)
Probability of successful eradication in a single time step given an infected area, A, and a maximum area considered for eradication, A_E	Binomial (1, $\frac{A}{A_E}$)	Binomial (1, $\frac{A}{A_E}$)
Economic		
Cost of eradication ($/ha)[i]	Pert (20000, 30000, 40000)	Pert (20000, 30000, 40000)
Demand elasticity[f]	Pert (−0.8, −0.6, −0.4)	Pert (−0.8, −0.6, −0.4)
Discount rate	0.08	0.08
Farm gate price of apples ($/$T$)	1,190 (i.e. p^*)	1,200 (i.e. p_0)[g]
Increased chemical cost ($/ha)[h]	1,000	1,000
Increased pruning expenses ($/ha)[h]	400	400

(*Continued*)

Table 6.1. (*Continued*)

Parameters	Quarantine Restricted Trade	Autarchic Trade
Maximum area considered for eradication (ha)	500	500
Retail marketing margin	Pert (0.3, 0.45, 0.6)	Pert (0.3, 0.45, 0.6)
Supply elasticity[f]	Pert (0.2, 0.5, 0.5)	Pert (0.2, 0.5, 0.5)
Volume imports from New Zealand (*T*)	2,650	0
Volume supplied by Australian producers (*T*)	Mean = 287,650(i.e. q_2)	290,300 (i.e. q_0)[d]
Wholesale marketing margin	Pert (0.1, 0.2, 0.3)	Pert (0.1, 0.2, 0.3)
Yield loss despite control (%)[h]	Pert (0, 10, 20)	Pert (0, 10, 20)

[a]Cook (2003).
[b]Paini *et al.* (2010).
[c]Waage *et al.* (2005).
[d]ABS (2004).
[e]Zadocks and Shein (1979), Waage *et al.* (2005).
[f]Valdes and Zietz (1980), Bhati and Rees (1996).
[g]HAL (2004).
[h]Cook (2003), McElwee (2000).
[i]Assumes zero compensation following tree removal, average planting density of 600 trees/ha and removal, transport, destruction and chemical costs of $16.70 per tree. This includes labour (team of three at $35/hr each), bulldozing equipment ($100/hr at 15 hours per ha), truck hire ($75/hr), incendiaries ($60/ha for green waste) and creation of a circular chemical buffer zone with a 500 m radius around previously infected sites. We assume a Bordeaux mixture (copper sulphate crystals and hydrated spay lime dissolved in water at a rate of 8:8:100) is used at a cost of $2.50/kg applied at 4 kg/ha in high water volumes to achieve a washing action on bark surfaces, taking approximately 2 hr per ha to apply).

attempt will successfully remove an infestation is proportional to the area the host plants occupied at the time of detection (see Table 6.1). If this does not occur before infection has spread to a predefined maximum area, A_E, which we have arbitrarily assumed is 500 hectares, the eradication attempt is aborted.

If detection does occur sufficiently early, eradication entails the complete removal of all infected trees and the creation of a chemical buffer zone around the area where the infection occurred. Infected trees are bulldozed and removed from quarantined properties by truck for incineration. This involves expenses of approximately $10,000 per hectare of trees removed. It is assumed that no compensation is offered to affected orchardists. After infected trees have been destroyed, the area is immediately replanted with apples and effectively re-enters production 6 years from the time of replanting. In present value terms, the cost of replanting one hectare of apples and waiting for the trees to mature to the point of yielding fruit is approximately $55,000.

When detection does not occur early in an outbreak or when an eradication attempt fails to prevent infection reaching the 500 hectare threshold level, eradication is aborted. This does not mean that fire blight now spreads unimpeded within the virtual world of the model since we assume that there are certain on-farm management strategies orchardists can adopt to reduce the impacts of the disease. For instance, aggressive pruning to remove infected spurs, shoots and limbs can greatly reduce fire blight incidence. Copper sprays can be applied in conjunction with the manual removal of fire blight cankers. Antibiotics can also be used to prevent fire blight infection on host flowers' stigma surfaces, although it is unlikely that they will be effective if applied more than 24 hr post infection so the timing of applications is critical. All of these techniques involve a cost to producers and are not guaranteed to be 100% effective. Yield losses despite control are estimated to average 10% per annum (represented in the model as Pert (0.0, 0.1, 0.2) (McElwee, 2000)).

The spread of fire blight is connected dynamically with the costs of eradication and on-farm control by simply multiplying the area infested by a constant marginal damage cost (or an average damage cost). For outbreaks involving less than 500 hectares, area is multiplied by eradication and replanting costs (see Table 6.1). When infection spreads beyond 500 hectares, the remaining area is multiplied by an average on-farm disease management cost comprising of additional chemical and application costs ($1,000 per hectare), pruning costs ($400 per hectare) and yield losses (Pert (0%, 10%, 20%) (McElwee, 2000; Cook, 2003)).

By summing the production losses over each time step and assuming fixed costs are zero, we estimate the present value of producer surplus losses attributable to fire blight incursions over a 30-year period. We do this for both an autarkic and a quarantine-restricted trade scenario by altering the parameter λ (i.e. the expected number of introductions per year), which is higher under the latter scenario, as explained earlier. The difference between estimated damages in each scenario indicates the change in expected impact of the disease over time as a result of granting NZ access to the Australian market for fresh apples, EI^*.

To determine the expected net gains from trade, NG_E, the traditional gains, GT (i.e. Equation (5.6)) must first be calculated. This requires additional model parameters as outlined in Section 5.2 and Table 6.1. The following parameters and data are used to represent them in the model:

Elasticity of Supply: Estimated using a pert distribution with a minimum value of 0.2, a most likely value of 0.5 and a maximum value of 0.8 (approximated using Valdes and Zietz (1980) and Bhati and Rees (1996) as guides).

Elasticity of Demand: Estimated using a pert distribution with a minimum value of −0.4, a most likely value of −0.6 and a maximum value of −0.8 (approximated using Valdes and Zietz (1980) and Bhati and Rees (1996) as guides).

Closed Economy Quantity Supplied (q_0): 290,300 T (HAL, 2004; ABS, 2004).[4]

Closed-Economy or Autarchic Price (p_0): $1,200/T (HAL, 2004).

Free Trade Price (p^{**}): This represents the price at which apples from external sources can be supplied to the Australian market without quarantine treatments. O'Rourke (2007) ranked Australia 16th in a study of the production efficiency of 28 of the world's apple-producing countries, while NZ was ranked second. Given the relatively close proximity of NZ apple producers to the Australian market, this suggests that the landed price of NZ apples is a reasonable proxy for the world price. Using figures from Cook (2008) as a guide, we assume that shipping costs (including loading and road freight costs) involved in transporting product from NZ to Australia are on the order of $400/T. Data from FAO (2009) from 2002 to 2006 were used to construct a linear time series to project a 2008 NZ farm gate price of $680/T. This implies a landed price of NZ apples in Australia of around $1,080/T.

Post-Quarantine or Import Price (p^*): In the absence of time-series price differentials in a closed and quarantine-restricted market, an approximate quarantine-induced price rise of 10% above the free trade price is assumed (i.e. $1,190/T), at which 2,650 T will be imported.

[4]The Australian Bureau of Statistics Catalogue No. 7121.0.55.002 is collected on behalf of the peak industry body, Apple and Pear Australia Limited.

Wholesale and Retail Marketing Margins: Wholesale margins are specified as a pert distribution with a minimum value of 10%, a most likely value of 20% and a maximum value of 30%. Retail margins are estimated using a pert distribution with a minimum value of 30%, a most likely value of 45% and a maximum value of 60% (Cook, 2008). Both wholesale and retail marketing margins are assumed constant in percentage terms.

6.2.3. *Results*

6.2.3.1. *Welfare analysis: Benefits from trade*

The parameterised model was used to calculate the net gains from trade resulting from the import of NZ apples (subject to quarantine treatments) into Australia. This represents a movement from a closed-economy situation (autarchy) to one of quarantine-restricted trade. Results are presented in Table 6.2.

Since the variability of parameters has been incorporated in the analysis, results are probability distributions rather than point estimates. They are described in Table 6.2 simply by using the mean and standard deviation. With the demand and supply curve specifications used in this exercise, the results indicate that the net gains from trade to the Australian economy from allowing NZ apples into the country subject to quarantine treatments (rather than excluding them completely) are expected to average around \$15.6 million per year with a standard deviation of approximately \$1.6 million.

Figure 6.1 shows the flow of these gains from trade over a 30-year period, showing the average present value ± 1 standard deviation from the mean. The discount rate used in these calculations is 8%. This relatively high rate consists of a margin of 3% on top of a real risk-free rate of 5% (Commonwealth of Australia, 2006).

Table 6.2. The gains from trade resulting from apple imports from New Zealand.

	Mean	Standard Deviation
Change in consumer surplus (ΔCS)	\$46,343,200	\$1,573,340
Change in producer surplus (ΔPS)	−\$30,731,670	\$400,150
Gains from trade (GT)	\$15,611,530	\$1,604,560

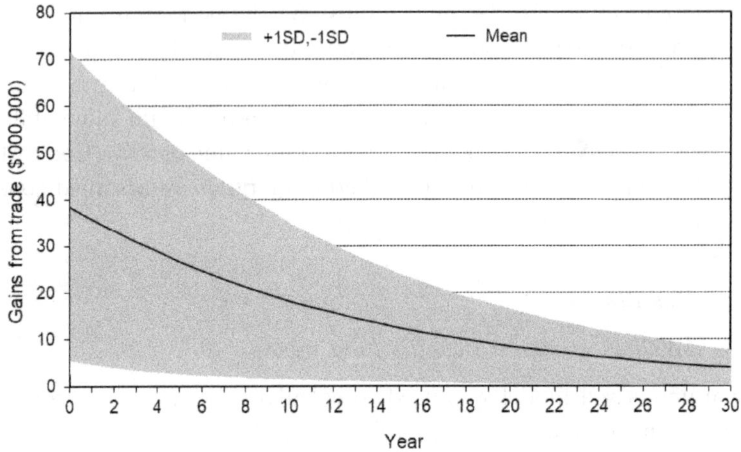

Figure 6.1. Present value of gains from trade from New Zealand apple imports over time.

The calculated gains from trade are responsive to changes in the landed price of NZ apples under both a free trade and quarantine-restricted scenario, as we would expect. All other things being equal, if freight and transport costs are as low as $300/T, the landed price of apples may be as low as $980/T. This would mean net gains from quarantine-restricted trade of $17.3 million, and therefore a higher break-even level of invasive species risk. On the other hand, if rising oil prices continue to increase the cost of freight to the point where they reach $600/T, the landed price of apples may be as high as $1,280/T. This would mean the cost of quarantine restrictions to Australian consumers would be negligible. Likewise, if SPS measures imposed on imported apples increase the landed price of NZ apples by more than 20% (above the base case level of $1,080/T), the gains to Australian consumers are negligible as the post-quarantine price would be equivalent to an autarkic price.

6.2.3.2. *Invasion costs*

The total invasion costs over a 30-year period were calculated for both the quarantine-restricted trade and autarchic trade, with the difference between them representing the change in expected impact. The present value of annual damages expected to result from fire blight incursions over the simulation period is given in Table 6.3. These results are obtained using the parameter values from Table 6.1.

Table 6.3. Expected damage per year from fire blight incursions.

	Mean	Standard Deviation
Expected impact under autarchy (EI_A)	\$17,282,810	\$14,242,790
Expected impact under quarantine-restricted Trade (EI_Q)	\$40,803,290	\$9,813,510
Change in expected impact (EI^*)	\$23,783,540	\$15,058,590

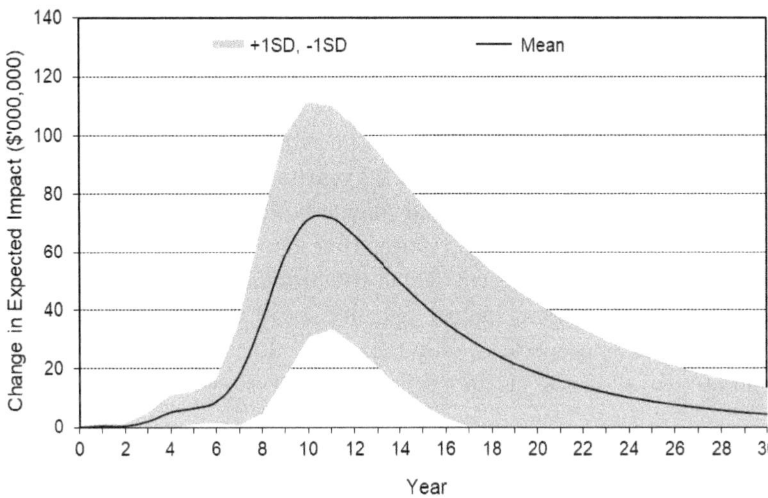

Figure 6.2. Increase of the expected net present value of fire blight damage costs in a scenario with quarantine-restricted trade with respect to a scenario with no trade. Impacts are composed of costs of eradication attempts, on-farm management efforts and market revenues losses.

The average present value from 5,000 iterations of the model ± 1 standard deviation from the mean is shown in Figure 6.2. The increase in expected losses from fire blight incursions will be relatively small until about year 6, after which they increase dramatically. This lag period is attributable to the SPS measures imposed on apple imports being relatively effective. Under the quarantine-restricted trade scenario, the model predicts an incursion event that is roughly four times more likely to occur than under an autarchy scenario. However, despite the difference in expected impact occurring well into the future (and hence being subject to the erosive effects of an 8% discount rate), the damage caused is large compared to consumer's gains.

6.2.3.3. *Expected net change in social welfare*

The overall welfare effects on the economy can be seen by plotting the change in expected impact with the gains from trade, as in Figure 6.3. Here the mean net change in social welfare represented by changes in consumer and producer surpluses under the autarchic and SPS-restricted trade scenarios is also plotted. Net social welfare is likely to be improved substantially by opening up trade to NZ apples for the first 7 years. Beyond this point, the increase in fire blight damages resulting from an increased likelihood of incursion will be such that net welfare is expected to become negative. This is despite the effects of discounting on future values using a high discount rate of 8%.

Given the accrual of benefits and costs over time, it is apparent that the success or failure of trade hinges on the time period being considered by decision makers. Assuming this to be 30 years and taking the mean present annual value of gains from trade and change expected impact resulting from a move from autarchy to SPS-restricted trade, the net impact on social welfare is likely to be negative. Table 6.4 presents cumulative gains from trade, change in expected impact and net social welfare change over 10-, 20- and 30-year time periods. Note that over 30 years, the ratio of trade benefits to expected costs is approximately 0.7:1. However, we should again

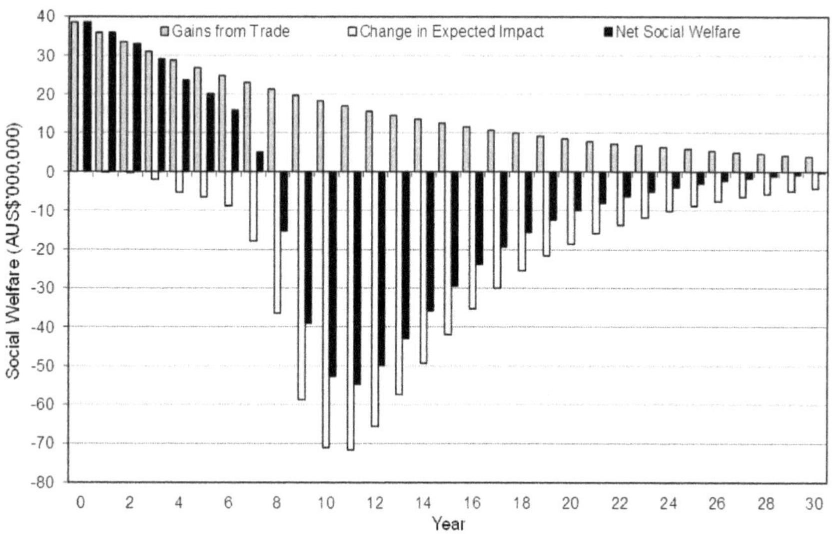

Figure 6.3. Change in social welfare over time resulting from New Zealand apple imports to Australia.

Table 6.4. Cumulative net change in social welfare from Australia importing apples from New Zealand.

Year	Gains from Trade — Australia ($ million)	Change in Expected Impact ($ million)	Net Change in Social Welfare ($ million)
10	301.3	207.1	94.2
20	424.0	624.0	−200.0
30	480.9	713.5	−232.6

point out the uncertainty in projecting this far into the future, as reflected in high standard deviations for both GT and EI^*, makes it difficult to draw definitive conclusions.

Note also that if we were to apply a personal discount rate of 10% to the mean or average model calculations, as opposed to a public/social discount rate of 8%, EI^* would fall to \$18.8 million. GT would also decrease to \$13.2 million, meaning that the ratio of trade benefits to expected costs would remain at 0.7:1. On the other hand, if we were to adopt a more precautionary approach to agricultural change and apply a discount rate of 5%, EI^* would increase to \$34.6 million and GT to \$20.7 million, reducing the ratio of benefits to costs to 0.6:1.

We have not considered the change in consumer and producer welfare in the exporting country in this example, but potentially the WTO could also take these into consideration as part of the DSP. In Section 5.3, we mentioned the likelihood that redirecting a portion of domestic production to an importing country would increase domestic prices, thereby decreasing the welfare of consumers in the exporting country. However, producer surplus in the exporting country is set to increase as the new market is opened up, so the gains from trade in the exporting country (i.e. GT_X from Equation (5.8)) depend on the relative sizes of consumer surplus and producer surplus changes (i.e. ΔCS_X and ΔPS_X, respectively).

This adds another layer of complexity to negotiations and underpinning analysis, but at the very least we can recognise that $GT - EI^*$ provides a breakeven requirement for GT_X to exceed. Recalling Equation (5.9), breakeven occurs where $NG_E = GT_T - EI^* = 0$. It may be that in a country like NZ with a relatively small number of consumers and a strong agricultural economy, ΔPS_X is large compared to ΔCS_X, in turn making GT_X large. In this case, it is possible that GT_X exceeds the breakeven level. If this is the case and the WTO were to take GT_X into consideration, it would be advantageous for trade to go ahead even if $GT - EI^* < 0$. Note that based on the results in Table 6.4, then GT_X over a 30-year period would need to exceed $232.6 million for this situation to apply.

6.3. Canadian Salmon Importation Request to Australia and the Risk of Whirling Disease Introduction

6.3.1. *Background*

In this section, we will look in detail at the Canadian salmon case outlined previously in Chapter 3. As mentioned in Chapter 3, Canada had sought access to the Australian salmon market since the imposition of quarantine restrictions on imported fish under the Australian Quarantine Proclamation of 1975. Canada claimed its products did not represent an excessive risk of disease transfer (Australian Quarantine and Inspection Service, 1995). The Australian restrictions prohibited the importation of salmon meat that had not been heat treated, implying that uncooked salmon meat represented an excessive level of disease risk. However, at the time there had been no attempts made to investigate the risks associated with uncooked salmon meat imports.

By 1995, annual production of Atlantic salmon in Australia had reached 7,000 tonnes, almost all of it produced in the island state of Tasmania. This made it the nation's second highest valued aquaculture industry with a gross value of $63 million (Industry Commission, 1996). Approximately 40% of industry revenue was generated by exports, mainly to Japan, while imports remained prohibited (Industry Commission, 1996). However, following GATT consultations in 1994, Australia finally agreed to undertake an Import Risk Analysis (IRA) to address the technical quarantine issues raised by the importation of wild, ocean-caught Pacific salmon from Canada (Thomson, 2001).

In May 1995, AQIS published a draft risk analysis which concluded that if certification conditions were met, the risk posed by imported salmon was negligible to Australian salmon producers (Australian Quarantine and Inspection Service, 1995). However, a final draft of this assessment published in December 1996 reported that as many as 20 exotic diseases could be introduced through imports, the economic consequences of which could be large (Stanton, 2001). On the basis of this final draft report, Australia's ban on imported salmon was upheld.

Canada's response was to launch a formal complaint through the WTO DSP in March 1997, requesting that the Dispute Settlement Body (DSB) establish a Panel to determine whether Australia's prohibition on uncooked salmon violated the SPS Agreement since it appeared to be based on a consequence assessment, as opposed to an IRA (Thomson, 2001). The Panel established to look into the issue in April 1997 handed down its final report in June 1998 in which it concluded that Australia had not demonstrated a rational relationship between the final risk assessment and import ban, and that the ban failed to comply with the SPS Agreement (WTO, 1998b).

Following Australia's inevitable appeal, the Appellate Body report released in October 1998 supported the view that the ban violated the terms and conditions of the SPS Agreement, finding that the IRA had not evaluated the likelihood of disease entry and establishment either with or without possible quarantine treatments (WTO, 1998a). It therefore represented a breach of Article 5 of the SPS Agreement, which deals with the assessment of risk and the determination of the appropriate level of SPS protection (GATT, 1994). The Panel also suggested that there were less trade-restrictive measures available to Australia than an import ban, but recognised the fact that the risk-abating properties of these measures were difficult to determine.

Although the time limit specified for Australia to comply with the DSB findings was 8 months after the adoption of the Panel and Appellate Body reports, Australia failed to meet this deadline until 2 weeks later (July, 1999). By this time, AQIS had completed yet another IRA, this time asserting that the importation of salmon meat from Canada 'under certain conditions' did not pose excessive risk to Australian fisheries (Australian Quarantine and Inspection Service, 1999).

Predictably, this drew the ire of Canada which requested WTO permission to take retaliatory action against Australia in July 1999. However, this action was delayed while the DSB considered the issue of Australia's compliance or non-compliance by way of the original Panel established in 1997. It subsequently ruled that this new risk assessment was compliant with the terms of the SPS Agreement, but that the suggested import conditions placed on imported salmon, whereby products were processed to a 'consumer-ready' form, were unnecessarily trade restrictive. This Panel report was formally adopted in March 2000 without appeal (Stanton, 2001).

In a side note, the decision of the Federal Government of Australia not to appeal the WTO ruling brought it into conflict with the Tasmanian State Government which felt such an action could severely impact its salmonid industry in the long term. It subsequently instigated its own ban in direct contravention of the Panel findings. Fortunately for other Australian states and industries, Canada did not take retaliatory actions.

6.3.2. *Modelling the supply curve shift in the event of whirling disease introduction*

In the analysis to follow, we will use a different spread model to the one presented in Section 6.2.2 as the area-based reaction diffusion model is unsuited to marine pests and diseases. Using the example of whirling disease (*Myxobolus cerebralis*), we will characterise its potential spread over time using the logistic model of Schaefer (1957) modified to allow us to specify the length of time it will take the disease to affect Australian salmon fisheries once introduced. This model is combined with a measure of the marginal damage cost to simulate costs related to whirling disease over a 30-year period under autarky and quarantine-restricted trade scenarios.

Unlike the fire blight case study, eradication is considered technically unachievable once the disease becomes established, and so it is not necessary to include eradication costs in the model. However, like the fire blight model, the whirling disease model relies on a Poisson stochastic process to simulate entry events over using the parameter λ to the expected number of introductions per time step.

We arbitrarily assume that, on average, one introduction event would occur every 10 years under an SPS-restricted trade scenario. However, this may be as high as one introduction event every 5 years, or as low as one every 15 years. We therefore define λ under SPS-restricted trade as Uniform $(0.07, 0.2)$. Under autarky, we assume λ falls to Uniform (0.01, 0.05).

In iterations of the model in which the Poisson process predicts a successful introduction, the probability of that event leading to successful establishment is assumed to be Uniform (0.7, 1.0) under both scenarios. The model assumes that a spatially homogenous environment exists through which whirling disease will spread once it becomes established. As it is deemed non-eradicable, it will continue to spread and become naturalised.

The area of infection is generated by the model of Schaefer (1957). We assume that after the establishment of the disease in Australia, the proportion of the total fish stock infected with whirling disease in period t $(S(t))$ increases over time following the logistic equation:

$$S(t) = S_{\max} \frac{I_{\max}}{1 + \left(\frac{I_{\max}}{I_{\min}} - 1\right) e^{-\omega t}} \tag{6.4}$$

Here, $S_{\max}$ is the total size of the fish stock, $I_{\max}$ is the maximum proportion of the fish stock infected, $I_{\min}$ is the proportion of the fish stock affected upon disease establishment and ω is infection growth rate.

In the absence of information about ω, we use a hypothetical infection growth determined by the number of time periods it takes the disease to affect a given proportion, θ, of $S_{\max}$ such that

$$\omega = -t^{\theta} \ln \left[\frac{I_{\max} - \theta}{\theta \left(\frac{I_{\max}}{I_{\min}} - 1 \right)} \right] \tag{6.5}$$

Here, θ is a specified proportion of $S_{\max}$ infected and t^{θ} is the number of periods taken for the disease to reach the infection level θ. The values and distributions assigned to each parameter of the whirling disease spread model are provided in Table 6.5.

Once it becomes established, the disease cannot be eradicated from a waterway due to the proliferation of intermediate host tubificid worms (Markiw and Wolf, 1983). If left untreated, it can cause exhaustion, malnutrition and deformities that reduce the proportion of stock that can be marketed. The severity of impact depends on the species of fish concerned, but generally juvenile salmon and trout show disease symptoms while adults rarely show any clinical signs when exposed to infection.

Experiences with whirling disease in Europe and the USA demonstrate that its effects can be managed in areas where it has become established. This requires fish to be reared in isolated, disease-free tanks before they are transferred to 'grow-out' facilities and diseased waters (McKelvie, 1991). Installation and operation of these tanks add substantially to the costs of production, especially given the fact that rearing tank water must be treated with ultraviolet irradiation and filtration systems not currently used in Australia.

We use estimated costs of whirling disease to the Tasmania salmonid industry reported in McKelvie (1991) to form a broad estimate of potential costs of a disease outbreak around the time of the Panel report, WTO (1998b). The production cost increase for a fishery supplying 1-year-old Atlantic salmon known as *smolt* was estimated at almost \$155,000 per year. This included purchase and installation of nine filtration units (\$75,000) and variable costs (i.e. electricity, repairs and maintenance) of \$80,000 (or \$35 per tonne) (McKelvie, 1991).

With three main salmon suppliers in Tasmania (Tassal, Huon Aquaculture and Petuna), fixed cost will increase by a total of \$225,000 across the

Table 6.5. Parameters and their assumed values.

Parameters	Quarantine-Restricted Trade	Autarchic Trade
Biological		
Infection growth, ω (unitless)[a]	2.26–6.79	2.26–6.79
Maximum proportion of fish stock infected, $I_{\max}$ (%)[a]	100	100
Minimum proportion of fish stock infected, $I_{\min}$ (%)[a]	0.01	0.01
Probability of entry, λ (%)[b]	Uniform (0.1, 5)	Uniform (0.0001, 0.1)
Probability of establishment (%)[b]	Uniform (70, 100)	Uniform (70, 100)
Proportion of fish stock infected at t^{θ}, θ (%)[a]	90	90
Time taken for θ of fish stock to be infected (yr)[a]	Uniform (8, 12)	Uniform (8, 12)
Economic		
Demand elasticity[c]	−0.28	−0.28
Discount rate[d]	0.08	0.08
Increased fixed production cost ($)[e]	225,000	225,000
Increased variable production cost ($/T)[e]	11	11
Price of salmon ($/T)[f]	9,500 (i.e. p^*)	10,000 (i.e. p_0)
Retail marketing margin	Pert (0.3, 0.45, 0.6)	Pert (0.3, 0.45, 0.6)
Supply elasticity[g]	1.45	1.45
Volume imports from Canada (T)	660	0
Volume supplied by Australian producers (T)[h]	7,650 (i.e. q_2)	7,090 (i.e. q_0)
Wholesale marketing margin	Pert (0.1, 0.2, 0.3)	Pert (0.1, 0.2, 0.3)
Yield loss despite control (%)	Pert (0, 10, 20)	Pert (0, 10, 20)

[a]See Equation (6.5) and explanation. The infection growth constant ω is determined by specifying a proportion of stock infected, θ, and the amount of time taken for this proportion to be reached (t^{θ}) once establishment has occurred. We assume it will take the disease 1–3 years to achieve an infection rate of 90%. This gives an infection growth parameter of 2.26–6.79.
[b]Assumes that the probability of entry in the quarantine-restricted trade and closed-economy scenarios correspond to the 'very low' and 'extremely low' risk categories outlined in Biosecurity Australia (2001), respectively. Introduction events are assumed to have a high probability of becoming established (between 70–100%) due to the abundance of intermediate hosts, aquatic tubificid oligochaete (Markiw and Wolf, 1983).
[c]Andersen, Roll and Tveteras (2008).
[d]Commonwealth of Australia (2006).
[e]McKelvie (1991).
[f]ABARE–BRS (2010).
[g]Seale, Regmi and Bernstein (2003).
[h]ADVS (1999).

industry if whirling disease becomes established. These costs are incurred immediately upon the disease becoming established.

In addition to these costs, export losses will be incurred when the disease becomes established. McKelvie (1991) considered losses of between \$2.4 million and \$14.5 million per year (i.e. 10–60% of total export revenue) possible if whirling disease led to the loss of the Japanese market. Although McKelvie (1991) used a lower estimate in their analysis, we simulate export market losses using a pert distribution with a minimum value of 10%, a most likely value of 35% and a maximum value of 60% (i.e. pert (0.1, 0.35, 0.6)).

Unlike fixed costs and export costs that take effect when the disease is first introduced, variable costs of management action increase as the proportion of total fish stock affected increases. Since this particular disease is expected to spread quickly, variable costs are assumed to peak between 1 and 3 years after introduction. They are calculated in the model by multiplying the proportion of fish stock affected by a constant marginal damage cost (i.e. \$35/tonne, as outlined previously).

Using the model to sum all costs over time, we estimate the present value of producer surplus losses attributable to whirling disease incursions over a 30-year period. We do this for both an autarkic and a quarantine-restricted trade scenario, and the difference between estimated damages in each scenario indicates the change in expected cost of the disease over time as a result of granting Canada access to the Australian market for salmon, EI^*.

The traditional gains from trade, GT (i.e. Equation (5.5)), are calculated using the same method outlined in Section 5.2. The necessary parameters are listed in Table 6.5. Note that the free trade price (i.e. p^{**} from Figure 5.1) is assumed to be \$9,000/T.

6.3.3. *Results*

6.3.3.1. *Welfare analysis: Benefits from trade*

The parameterised model was used to calculate the net gains from trade resulting from the import of Canadian salmon subject to quarantine treatments into Australia. Given the historical ban on salmon products entering Australia prior to 1998, this constitutes a change from a closed-economy situation to one of quarantine-restricted trade. Results are presented in Table 6.6.

Table 6.6. The gains from trade resulting from salmon imports from Canada.

	Mean	Standard Deviation
Change in consumer surplus (ΔCS)	\$6,699,870	\$334,580
Change in producer surplus (ΔPS)	−\$3,573,350	\$0
Gains from trade (GT)	\$3,126,520	\$334,580

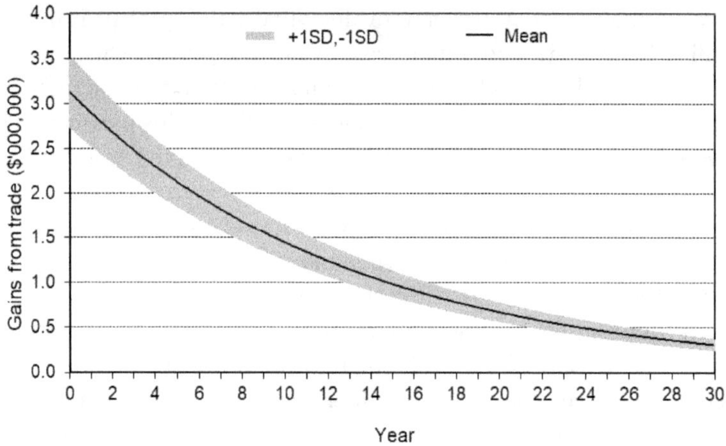

Figure 6.4. Present value of gains from trade from Canadian salmon imports over time.

Most parameters are specified as point estimates in this analysis. As there is no variability in parameters used to calculate producer surplus changes, the standard deviation of results is zero. However, marketing margins are entered as distributions, so there is variability in the calculation of consumer surplus. Results in the table indicate that the net gains from trade to the Australian economy from allowing Canadian salmon imports into the country subject to quarantine treatments are likely to be \$3.1 million per year with a standard deviation of \$0.3 million. Figure 6.4 plots these gains over a 30-year period, showing the average present value ±1 standard deviation from the mean. The discount rate used in these calculations is the same as the NZ apple case study, 8%.

6.3.3.2. *Invasion costs*

The total costs over a 30-year period were calculated for both the quarantine-restricted trade and closed-economy scenario, and then the

Table 6.7. Expected damage per year from whirling disease incursions.

	Mean	Standard Deviation[a]
Expected impact under autarchy (EI_A)	\$157,260	\$1,080,210
Expected impact under quarantine-restricted trade (EI_Q)	\$317,190	\$2,060,100
Change in expected impact (EI^*)	\$159,930	\$1,010,990

[a]Large standard deviations result from the use of broad entry probability distributions of Uniform (0.1%, 5%) and Uniform (0.0001%, 0.1%) for the quarantine-restricted trade and closed-economy scenarios, respectively. These are taken from the semi-quantitative risk analysis methodology outlined in Biosecurity Australia (2001) and relate to 'very low' and 'extremely low' risk categories, respectively.

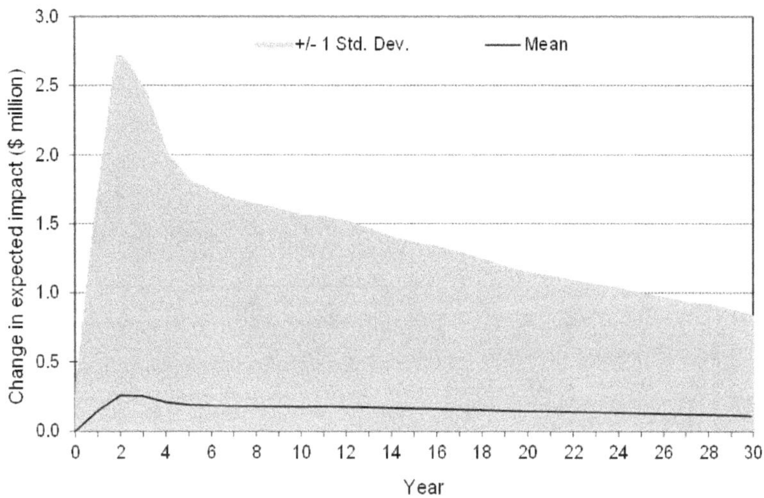

Figure 6.5. Increase of the expected net present value of whirling disease damage costs in a scenario with quarantine-restricted trade with respect to a scenario of no trade.

difference between them calculated to reveal the change in disease impact likely to result by allowing salmon imports from Canada. Using the parameter values from Table 6.5, the present value of annual damages expected to result from whirling disease incursions over the simulation period is given in Table 6.7.

The average present value from 10,000 iterations of the model ± 1 standard deviation from the mean is shown in Figure 6.5. Average losses predicted by the model are relatively small (i.e. <\$0.5 million per year), with the remoteness of Tasmanian production facilities and SPS measures making entry events extremely rare.

6.3.3.3. *Expected net change in social welfare*

The overall welfare effects on the economy can be seen by plotting the change in expected impact with the gains from trade, as in Figure 6.6. Here the mean net change in social welfare represented by changes in consumer and producer surpluses under the closed-economy and SPS-restricted trade scenarios is also plotted. Net social welfare is likely to be improved in the short term by importing Canadian salmon, but after 5–6 years the increase in expected whirling disease damage is likely to outweigh the gains from trade. This is not to say that the damages likely to result from the disease are particularly large, but rather the gains from trade are relatively small.

As we noted in the previous case study, the net social welfare benefit resulting from trade depends on the period being considered by decision makers. Table 6.8 shows the cumulative gains from trade, expected impact and net social welfare summed over 10-, 20- and 30-year periods. If we are judging the decision to relax the trade ban on Canadian

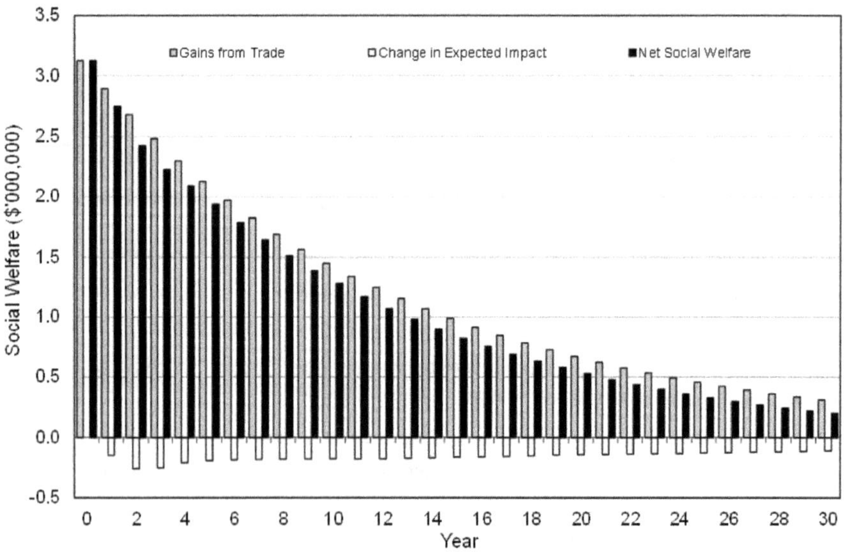

Figure 6.6. Change in social welfare over time resulting from Canadian salmon imports to Australia.

Table 6.8. Cumulative net change in social welfare from Australia importing salmon from Canada.

Year	Gains from Trade — Australia ($ million)	Change in Expected Impact ($ million)	Net Change in Social Welfare ($ million)
10	24.1	−2.0	22.1
20	33.8	−3.6	30.2
30	38.3	−4.8	33.5

salmon over 30 years, the ratio of trade benefits to expected costs is approximately 8:1.

Finally, and again in contrast to the previous case study, note that any consideration of gains from trade for the exporting country would in this case only enhance the estimated net social welfare benefit from allowing imports of fresh Canadian salmon into Australia.

6.4. Conclusion

This chapter has applied the methodology developed in Chapter 5 by combining gains from trade with the impact of trade on biosecurity risk in the importing country to provide an evaluation of the impact of trade on social welfare. Applying this methodology to case studies of the Australian bans on NZ apples and Canadian fresh salmon as outlined in Chapter 3, we have demonstrated in broad terms how such a methodology might potentially be used in a WTO DSP to evaluate which trade policies are likely to lead to the best social welfare outcomes.

While the quantification of these case studies has been largely illustrative, this chapter has demonstrated in broad terms how the methodology of Chapter 5 can lead to different dispute resolutions than that produced by the existing rules based approach of the WTO. Specifically, in the case of NZ apples, the expected disease outbreak costs to domestic production associated with imports were larger than the estimated gains from allowing imports, implying that allowing trade results in a net loss of social welfare and thereby contradicting the WTO ruling. In the Canadian salmon case however, the predicted gains from trade were larger than the expected disease costs, implying that allowing trade results in a net gain in social welfare and thereby endorsing the WTO ruling. On this basis, we argue that economic frameworks where the benefits of trade to consumers and bioeconomic modelling of invasive species' spread and control are reconciled are needed to solve import trade disputes in the best interest of social welfare.

In Chapter 7, we will discuss extensions to this social welfare evaluation approach to incorporate an analysis of ecosystem service changes that could potentially result from the production and trade of goods.

Part C
Evaluating International Trade Proposals in the Context of Ecosystem Services Risks

Chapter 7

Evaluating International Trade Proposals in the Context of Ecosystem Services Risks: Methodology

7.1. Introduction

In this chapter, we concentrate on ecosystem services other than those that directly provide services relating to food and water which have been dealt with under biosecurity in Chapters 5 and 6. Rather, we concentrate here on those aspects of ecosystem services which have value to people but which are not traded directly through markets.

As outlined in Chapter 4, *ecosystem services* as a concept helps synthesise essential social, environmental and economic concepts, allowing researchers and managers to link human and ecological systems in a viable and policy-relevant manner (Liu *et al.*, 2010a). Ecosystem services are the benefits provided by ecosystems, including provisioning services such as food and water; regulating services such as regulation of floods, drought, land degradation and disease; and cultural services such as recreational, spiritual, religious and other non-material benefits (Costanza *et al.*, 1997; Millennium Ecosystem Assessment, 2005). As such, the risks of ecosystem services being altered as a result of changing patterns of trade are multifarious (Ekins *et al.*, 1994; Koellner, 2011). Nevertheless, they could be taken into account in trade negotiations if included in the same way, invasive species risks have been included in the evaluation methodology developed in Chapter 5 and illustrated in Chapter 6.

However, as demonstrated in Chapter 4, the WTO has continued to approach trade–ecosystem services disputes through interpretation of Article XX with scant attention given to trade, economic or environmental implications.

More specifically, it was found in Chapter 4 that the core problem is the potential inability for the WTO's rules-based approach to deliver against its organisational objective of maximising social welfare. In particular, this rules-based approach as it stands considers only the 'intent' of an ecosystem services action not its impact or 'effect'.

As a consequence of these limitations, this chapter develops a conceptual framework on the basis of which trade dispute decisions that account both of the potential gains from trade and of the associated ecosystem services impacts can be made, and in so doing enable decisions which improve social welfare.

The structure of the chapter is as follows: Section 7.2 develops the methodological framework for incorporating ecosystem services impacts, as well as the impacts on consumer and producer surplus, into the total evaluation of the impacts of trade; Section 7.3 analyses the range of possible outcomes from this evaluation which takes account of both the gains from trade impacts and the ecosystem services impacts; and Section 7.4 provides some discussion of this analysis, while Section 7.5 concludes the chapter by summarising its contents and leading into the two case study applications of this methodology presented in Chapter 8.

7.2. A Framework to Determine Net Changes in Ecosystem Services

Sanitary and phytosanitary methods are applied to trade agreements to protect human, animal and plant life or health from risks arising from the introduction or spread of invasive species. They are subject to rules set under the SPS Agreement. All WTO member countries are signatories to this agreement, under which they have both rights and obligations.

Trade and market access divisions of member governments are responsible for representing national interests in the international arena. Each country is entitled to maintain a level of import security it considers appropriate to protect the welfare of individuals and communities within its territory, which is called the appropriate level of protection (ALOP). The basic obligations are that SPS measures must (i) be applied to the extent necessary to protect life or health and not be more trade-restrictive than required, (ii) be based on scientific principles and not maintained without sufficient scientific evidence, and (iii) not constitute arbitrary or unjustified treatment or be a disguised restriction on trade.

Governments of importing countries are responsible for conducting science-based risk assessments to determine appropriate conditions of entry for specific products based on their ALOP. Assume that a country is confronted with the choice of whether or not to import a homogenous good from an overseas supplier. Further assume that the landed price of the imported product (p_1) is below that of a domestic equivalent (p_0). The domestic market for the product is characterised by a downward sloping demand curve, $f(q)$, and an upward sloping supply curve, $g(q)$, shown in the left panel of Figure 7.1.

The left-hand panel depicts the market for a good in a country that is considering importing the same good from overseas suppliers. Here, the demand schedule, $f(q)$, and supply schedule, $g(q)$, intersect at price p_0, which is above the potential landed price for imports (p_1). The right-hand panel depicts the market for the good in the country wishing to export, where the demand schedule, $f^*(q)$, and supply schedule, $g^*(q)$, intersect at price p_2. Hence, traditional gains from trade will result if the two countries commence trade.

The curves $h(q)$ and $h^*(q)$ show the supply curves in the importing and exporting countries, respectively, inclusive of ecosystem services impacted in the supply process. Increases in ecosystem services damage in the

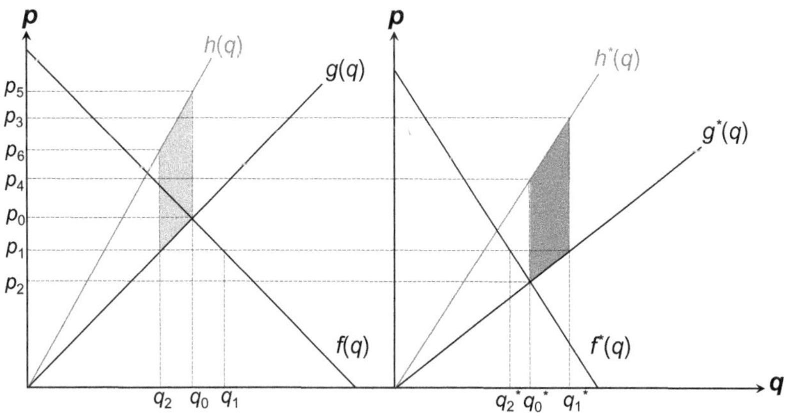

Figure 7.1. Trade decisions and effects on ecosystem services.

exporting country (represented as the heavily shaded region in the right-hand panel) must be weighed against both prevented ecosystem services losses in the importing country (shown as the lightly shaded region in the left-hand panel) and the conventional gains from trade for the complete net welfare effect of trade to become clear.[1]

Given this specification, we can measure changes in market-based welfare using consumer and producer surplus. Producer surplus is defined as net revenue earned by a producer from the sale of a good at a price above the minimum acceptable price they would have been willing to sell for before having to leave the market. Consumer surplus is the financial equivalent of the extra satisfaction (or utility) gained by consumers from purchasing a good at a price lower than what they were willing to pay for it.

Firstly, consider the effects of trade in the domestic market for the commodity in the importing country. If the decision was made to allow foreign product in it would favour consumers who would have access to a greater quantity of the good at a cheaper price. Rather than having to pay their domestic producers p_0, they can now pay p_1. The quantity they demand increases accordingly from q_0 to q_0. The welfare they gain as a result can be determined as

$$\Delta CS = \left[\int_0^{q_0} f^{-1}(q) - p_0\right] \mathrm{d}q - \left[\int_0^{q_1} f^{-1}(q) - p_1\right] \mathrm{d}q \tag{7.1}$$

While consumers gain, domestic producers who compete with the cheaper imported good lose out as they are losing market share. At the new lower market price p_1 they are only willing to supply q_2. Their total welfare loss can be measured as

$$\Delta PS = \left[p_0 - \int_0^{q_0} g^{-1}(q)\right] \mathrm{d}q - \left[p_1 - \int_0^{q_2} g^{-1}(q)\right] \mathrm{d}q \tag{7.2}$$

[1]Note that this representation is based on specifying the (negative) ecosystem services impacts as positively related to the level of production in each country, i.e. such that increased domestic production causes increased negative ecosystem services impacts in the domestic economy. Such a specification is consistent with a situation where domestic production damages the domestic natural environment. However, an alternative specification would feature the imported good damaging ecosystem services in the importing country, such as would occur in the situation where the imported good damaged the health of citizens of the importing country. Both these alternative specifications are illustrated by the case studies in Chapter 8.

The total net change in consumer and producer surplus or the traditional gains from trade in the importing country (GT_{T}) are therefore

$$GT_{\mathrm{T}} = \Delta PS + \Delta CS \tag{7.3}$$

Now, consider the effects of trade in the market for the traded good in the exporting country. Assume that it is characterised by the demand curve $f^*(q)$ and short-run supply curve $g^*(q)$ in the right-hand panel of Figure 7.1. The market clearing price in the absence of trade is p_2 (where consumers demand q_0^*, which is below the price at which it supplies the market in the importing country, p_1). The quantity demanded by exporting-country consumers at this higher price is q_2^*, leaving the quantity $q_1^* - q_2^*$ available for export (i.e. $q_1^* - q_2^* = q_1 - q_2$). Exporting-country consumer surplus, CS^*, will decline since $q_1 > q_2$.

$$\Delta CS^* = \left[\int_0^{q_0^*} f^{*-1}(q) - p_2\right] \mathrm{d}q - \left[\int_0^{q_2^*} f^{*-1}(q) - p_1\right] \mathrm{d}q \tag{7.4}$$

This negative consumer effect is offset by a rise in exporting-country producer surplus, PS^*. Producers are now supplying a greater quantity at a higher price.

$$\Delta PS^* = \left[p_1 - \int_0^{q_1^*} g^{*-1}(q)\right] \mathrm{d}q - \left[p_2 - \int_0^{q_0^*} g^{*-1}(q)\right] \mathrm{d}q \tag{7.5}$$

The traditional gains from trade for the exporting economy (GT_{T}^*) can therefore be determined as

$$GT_{\mathrm{T}}^* = \Delta PS^* + \Delta CS^* \tag{7.6}$$

If these gains were to be considered by the importing country, multilateral gains from trade (GT_{M}) can then be determined as

$$GT_{\mathrm{M}} = GT_{\mathrm{T}} + GT_{\mathrm{T}}^* \tag{7.7}$$

The implication is that the importing country would positively influence the social welfare of the global economy by entering into trade as long as $GT_{\mathrm{M}} > 0$.

As the primary institution overseeing international trade, the WTO plays a vital role in helping trading nations to align their knowledge and expectations of trade opportunities (Langlois and Foss, 1999). If discussions between prospective trading partners facilitated by the WTO were to involve information on both GT_{T} and GT_{T}^*, trade decisions could be made on the basis of potential net changes in global consumer and producer welfare (Cook and Fraser, 2008). This would require sufficient scientific and economic information about the size and distribution of intertemporal trade benefits and costs.

At this point, the limitations of consumer surplus and producer surplus as welfare measures need to be considered. In particular, their reliance on the price mechanism to determine who wins and who loses out from trade is restrictive since it does not account for non-market impacts. This means that environmental externalities resulting from the production of the good both in the importing and exporting countries are not accounted for in the supply curves of Figure 7.1.

There are several reasons for this. Firstly, environmental externalities resulting from production decisions in both countries will most probably involve changes in the abundance and distribution of an environmental resource, rather than its complete removal. Establishing definitive causes and effects often makes these changes difficult to measure, and therefore, to value. Secondly, the willingness of a person to pay to protect an environmental good is determined by many sociological elements involved in their decision-making process, such as age, income, upbringing, health, education, number of dependents, and so forth (Whitby, 2000). Including this information for all those affected by an environmental management decision is impractical. Finally, non-use values for environmental amenities must be considered, particularly when a change induced by a trade decision may have irreversible environmental consequences. Non-use values imply anyone can derive welfare from an environmental good irrespective of their intention or ability to physically use it.

For the purposes of a one-off trade decision this complex information is usually ignored since it is simply too difficult to collect and interpret. However, instead of valuing the state of environmental assets, it is possible to link agricultural land-use decisions with the value of services provided by the surrounding environment. In relation to the traditional framework for evaluating multilateral gains from trade outlined above, this framework can be seen to capture the benefits of the market-based provisioning services provided by ecosystems, but no account is taken of the non-market impacts of trade on regulating and cultural services.

However, if global trade approval decisions were to consider non-market ecosystem services changes in addition to GT_{M} in their assessment of social welfare changes, there could be sizeable implications. Returning to Figure 7.1, $h(q)$ and $h^{*}(q)$ represent the supply curves with ecosystem services changes embedded within them for the importing and the exporting country, respectively. As they have been depicted, any positive level of production carries with it a negative impact on ecosystem services that effectively moves the supply curves towards the vertical axis.[2] This reflects the fact that if social welfare includes ecosystem services impacts, each unit of production carries with it additional costs. This still omits impacts that are not reflected in changes in producer and consumer surplus. These include many non-use values. These changes in welfare may be significant and should be included, but for the purposes of this discussion, assume that they are fully captured in changes in producer and consumer surplus.

The decision to allow goods to be imported can now include at least a subset of ecosystem services lost as a result of increasing agricultural output (ΔES^*). In the absence of trade, q_0^* is supplied by producers in the overseas country for their domestic market. If the ecosystem services impacts of production are included in the price, they will charge consumers p_4. If trade commences, the quantity supplied by the overseas producers to both markets increases to q_1^*. At this level of production, the price would be p_3 if it were to include ecosystem services effects. Hence, ΔES^* can be determined as

$$\Delta ES^* = \left[\int_0^{q_1^*} h^{*-1}(q) - p_3\right] \mathrm{d}q - \left[\int_0^{q_{0*}} h^{*-1}(q) - p_4\right] \mathrm{d}q \tag{7.8}$$

The lost ecosystem services in the exporting country are sometimes referred to as 'leakage' (Engel *et al.*, 2008). They are offset by those gained in the importing country (ΔES) as a result of agricultural production contracting from q_0 to q_2 under competitive pressure from imports:

$$\Delta ES = \left[\int_0^{q_0^*} h^{-1}(q) - p_5\right] \mathrm{d}q - \left[\int_0^{q_2} h^{-1}(q) - p_6\right] \mathrm{d}q \tag{7.9}$$

The multilateral change in ecosystem services (ES_M) can therefore be determined as

$$ES_{\mathrm{M}} = \Delta ES = \Delta ES^* \tag{7.10}$$

[2]As stated earlier, an alternative specification would feature the imported good damaging ecosystem services in the importing country, such as would occur in a situation where the imported good damaged the health of citizens of the importing country. Both these alternative specifications are illustrated by the case studies in Chapter 8.

The net social welfare result of the decision to commence trade (W) is

$$W = GT_{\mathrm{M}} + ES_{\mathrm{M}} \tag{7.11}$$

7.3. Analysis

Equation (7.11) provides an expanded framework for enabling a global entity such as the WTO to make trade approval decisions based on potential net changes in multilateral social welfare and where this measure of welfare incorporates not just the traditional market net benefits of trade in goods and services (i.e. the joint changes in the economic measures of consumer and producer surplus) but also the non-market net benefits of changes in regulating services and cultural services (i.e. the joint environmental and social consequences of allowing trade).

Given this framework, there are six possible cases to consider:

(1) $GT_{\mathrm{M}} > 0$ and $ES_{\mathrm{M}} > 0$. This case is in effect a 'win–win' with unambiguous multilateral market and non-market welfare gains. In this case, trade would have been approved in the absence of considering changes in non-market ecosystem services, but the inclusion of such changes only reinforces the judgement to allow trade.

(2) $GT_{\mathrm{M}} < 0$ and $ES_{\mathrm{M}} < 0$. This case is in effect a 'lose–lose' with unambiguous multilateral market and non-market welfare losses. The judgement not to allow trade based just on the market gains from trade is reinforced by the additional consideration of changes in non-market ecosystem services.

(3) $GT_{\mathrm{M}} > 0, ES_{\mathrm{M}} < 0$ and $GT_{\mathrm{M}} + ES_{\mathrm{M}} > 0$. In this case, the traditional market gains from trade are undermined by losses in non-market ecosystem services. Nevertheless, there is an overall multilateral welfare benefit from allowing trade because the market benefits more than offset the non-market costs.

(4) $GT_{\mathrm{M}} > 0, ES_{\mathrm{M}} < 0$ and $GT_{\mathrm{M}} + ES_{\mathrm{M}} < 0$. In this case, there is a contrast between market benefits and non-market costs which highlights the difference between the traditional and our expanded framework for trade approval decision-making. Specifically, in this case, the multilateral costs to non-market ecosystem services outweigh the multilateral market benefits and trade should not be allowed.

(5) $GT_{\mathrm{M}} < 0, ES_{\mathrm{M}} > 0$ and $GT_{\mathrm{M}} + ES_{\mathrm{M}} < 0$. In this case, there are net losses in multilateral consumer and producer surpluses, which in the traditional framework implies trade would not be allowed. Although there are multilateral gains in non-market ecosystem services from trade taking place, these are not sufficient to offset the market losses and so trade would continue not to be allowed even with our expanded decision-making framework.

(6) $GT_{\mathrm{M}} < 0, ES_{\mathrm{M}} > 0$ and $GT_{\mathrm{M}} + ES_{\mathrm{M}} > 0$. As with Case (4), this case also represents a situation where there is a difference between the trade decisions made using the traditional and our expanded frameworks. Specifically, trade is not justified based just on changes in multilateral consumer and producer surpluses, but the gains in multilateral non-market ecosystem services from trade are sufficient to offset these market losses and so trade should be allowed.

In summary, of the six cases, four of them reveal a consistency between the trade approval decision made simply on the basis of multilateral market gains or losses and that made using our expanded framework which takes account of both multilateral market impacts and multilateral non-market ecosystem services impacts. However, there are also two cases (i.e. Cases (4) and (6)) where the trade approval decision based just on market impacts would be over-turned by also taking account of the non-market ecosystem services impacts. Moreover, in one of these cases, trade that would have been approved is no longer because of the dominating negative consequences of trade for non-market ecosystem services, while in the other case, the trade approval decision is reversed from not approved to approved because of the dominating positive consequences of trade for non-market ecosystem services.

Finally, these two special cases may be connected to the relationships between $h(q)$ and $g(q)$ (i.e. the importing country's traditional and expanded framework supply curves) and between $h^*(q)$ and $g^*(q)$ (i.e. the exporting country's traditional and expanded frameworks supply curves) in Figure 7.1.

Specifically, Case (4) would represent a situation where $h^*(q)$ was substantially above $g^*(q)$ by comparison with $h(q)$ and $g(q)$, so that the non-market ecosystem services costs of expanding production in the exporting country considerably outweigh the non-market ecosystem benefits from decreasing production in the importing country. Alternatively, Case (6) would represent a situation where $h(q)$ was substantially above $g(q)$ by comparison with $h^*(q)$ and $g^*(q)$ so that the non-market ecosystem services benefits of reducing production in the importing country considerably outweigh the non-market ecosystem services costs of expanding production in the exporting country.

7.4. Discussion

While literature illustrating the types of ecosystem services affected by trade is increasing, the same cannot be said of economic research focused on how to deal with these problems at an international level. There has been a historical tendency for benefit cost analyses of agricultural trade to focus on potential impacts of invasive species on industries within importing countries, rather than systemic effects (Perrings *et al.*, 2000). For instance, Costello and McAusland (2003) and Horan *et al.* (2002b) feature losses related to agricultural production most prominently but ignore environmental impacts. There have been several notable studies focusing on interactions between trade, invaders and native species (Barbier *et al.*, 2002; Finnoff and Tschirhart, 2005; Gutierrez and Regev, 2005; Knowler, 2005). However, these have not involved comparisons of ecosystem services effects.

This is not surprising given the lack of analytical support systems for trade decisions incorporating net ecosystem services effects. They require formal scientific knowledge to be combined with highly specific data capturing local contexts to determine net social welfare outcomes from trade in different products. The costs of gathering and interpreting this information too are large as are time lags involved in these processes.

Analytical support systems could be encouraged to develop through the adoption of flexible data collection and analysis standards. This might include qualitative information collection and interpretation in addition to more traditional quantitative information about the effects of trade on ecosystem services, and possibly local information that can be contextualised across a range of different disciplines (Nelson *et al.*, 2008). Scientific effort could also be redirected from attempts to achieve objectivity through uniform measures of ecosystem performance to provide more detailed interpretive measures of the impact of trade on social welfare. These might include, for instance, maintaining the greatest variety of distinct plant and animal lineages, preserving spectacular wild landscapes or protecting nature in a way that provides for the wellbeing of people living alongside conservation areas (Kareiva and Marvier, 2003).

Given the complex mixture of uncertain market and non-monetised non-market trade impacts, multi-criteria decision analysis (MCDA) techniques may offer a practical solution to trade policy-makers by stimulating discussion about possible policy choices, trade-offs and uncertainties. MCDA identifies solutions to complex and uncertain decisions characterised by alternatives/options that can be evaluated using weighable criteria (Jankowski and Nyerges, 2001). The weights provide a transparent bias which can be manipulated according to the idiosyncrasies of the decision or site affected by a decision. For the weighting process to function effectively, decision makers should have access to a wide range of information and decision support tools so that they may arrive at an appropriate set of weights through a process of consultation and deliberation (Burgman *et al.*, 2006). In the context of a decision of whether to allow trade in a particular commodity or not, the relative performance of each option (i.e. trade, no trade, limited trade) against the weighted criteria set allows policymakers to identify the optimal alternatives. In the broader context of global trading, life-cycle assessments and green national accounts, two widely used accounting tools, could be incorporated to increase trading nations' awareness of the use and the limits of ecosystems (Koellner *et al.*, 2011; Weber, 2011). These potential solutions to the problems of complexity, uncertainty and non-monetised values in trade dispute resolution situations are considered in more detail in Chapter 9.

7.5. Conclusion

This chapter has been motivated by the criticism in Chapter 4 that the core problem of the WTO's rules-based approach to trade dispute resolution in situations involving ecosystem services impacts is the potential inability for this approach to deliver against its organisational objective of maximising social welfare. In particular, this rules-based approach as it stands considers only 'intent' of an ecosystem services action not its impact or 'effect'. In an attempt to take account of these limitations, this chapter has developed a conceptual framework on the basis of which trade dispute decisions that account both of the potential gains from trade and of the associated ecosystem services impacts can be made, and in so doing enable decisions which improve social welfare.

In Chapter 8, we apply this methodology to two of the ecosystem services-related trade dispute cases identified and discussed in Chapter 4 in order to illustrate how the methodology can be used to improve the basis on which trade dispute decisions are made.

However, given the complexity and uncertainty associated with evaluating non-market ecosystem services impacts, we consider that a flexible approach to the collection and interpretation of ecosystem services impact data may be needed. As a consequence, in Chapter 9, we discuss the potential for techniques like MCDA to be used to facilitate trade decisions.

Chapter 8

Evaluating International Trade Proposals in the Context of Ecosystem Services Risks: Case Studies

8.1. Introduction

In this chapter, we select some of the case studies presented in Chapter 4 to illustrate application of the methodology developed in Chapter 7 to evaluate producer, consumer and ecosystem services gains and losses from trade. The first case study involves retreaded tyres imported to Brazil, and the second tuna imported to the US from Mexico. Recall that in both cases, the production of the goods concerned causes negative environmental and human health externalities. Following the methodology set out in Chapter 7, we use the traditional gains from trade as a benchmark with which to compare ecosystem services losses resulting from trade to indicate if trade is likely to be welfare enhancing or retarding.

Note, however, that instead of trying to directly estimate the value of ecosystem services losses avoided by banning trade in a particular product, this chapter first estimates using the dynamic model outlined in Chapter 6 just the (negative) market-based impact of the trade ban for each case study. Then, for each case study, we develop illustrative estimates of the loss in ecosystem services avoided by banning trade using the traditional avoided cost methodology. The net loss created by banning trade is then compared with these estimates of the loss in ecosystem services avoided by banning trade, including a break-even analysis of the results as a form of sensitivity analysis. However, Chapter 9 considers solutions to the evaluation complexities introduced by both estimation uncertainty and non-monetised values which would allow the direct comparison of market-based and ecosystem services impacts of banning trade.

8.2. Measures Affecting Brazilian Importation of Retreaded Tyres

8.2.1. *Reviewing the issue*

As discussed in Chapter 4, retreaded tyres are used tyres from passenger and commercial vehicles that have been reconditioned for further use. Reconditioning involves the removal of the old tread from a worn tyre and reclothing the original casting with a tread made from new materials (van Beukering and Janssen, 2001). This can be done via several different methods, including top-capping, recapping and remoulding (Van Damme, 2008), but we do not differentiate between them and just treat retreaded tyres as a homogenous product.

A country with a functional domestic retreading industry receives ecosystem services benefits by saving 80% of raw material and energy necessary for production of new tyres (van Beukering and Janssen, 2001). If it can export retreaded tyres, it will also receive ecosystem services benefits in the form of reduced pollution from discarded tyre carcases.

The downside is that the lifespan of retreaded tyres is generally less than new tyres, and under most circumstances, tyres can only be retreaded once. This means that the volume of waste tyres can be greatly increased through a reliance on imported retreads. Therefore, while the exporting country receives payment for its retreaded tyres, the problem of disposing of them once used is shifted to the importing country.

Additional human health problems for countries importing retreaded tyres can also occur, particularly if, at the end of their life, tyres are disposed of in tropical climates favourable to mosquitoes. Tyres stored in landfills or disposed of in illegal dumps tend to accumulate water and easily become vectors for diseases such as yellow fever, malaria and dengue fever. The storage and disposal of used tyres, which increases with importation of used and retreaded tyres, is thus likely to have adverse effects on human health and the environment (ICTSD, 2010).

Despite having a large tyre retreading industry of its own, Brazil is a net importer of retreaded tyres. Between 2001 and 2005, Brazilian producers collected and retreaded 55-million car tyres (WTO, 2007). In 1999, Brazil imported 18,455 tonnes of retreaded tyres (WTO, 2007), almost 70% of which were imported from the European Union (EU) (Brown and Trachtman, 2009).

Brazil imposed a ban on imports of retreaded tyres in September 2000, arguing such a ban was necessary to prevent risks to human, animal and plant life and health arising from the accumulation of waste tyres (Brown and Trachtman, 2009; Gray, 2008). The ban initially applied in a non-discriminatory fashion to imports from all foreign sources. Two independent dispute settlement proceedings to two different international organisations affected by the ban were subsequently triggered.

In the first case, Uruguay began proceedings against Brazil under the auspices of the Mercado Común del Sur (Southern Common Market, henceforth 'Mercosur'). In January 2002, a Mercosur ruling found the ban to be in violation of Mercosur law (Lavranos and Vielliard, 2008). Brazil complied with the ruling by modifying the ban to exempt imports of retreaded tyres from Mercosur-member countries Uruguay, Paraguay and Argentina, while non-Mercosur members were still affected by the ban (Brown and Trachtman, 2009).

In the second case, the European retread industry's trade association, the Bureau international permanent des associations de vendeurs et rechapeurs de pneumatiques (BIPAVER), initiated a complaint under the European Commission Trade Barriers Regulation in November 2003. This complaint challenged the Brazilian import ban as applied to imports from non-Mercosur countries, and led to the EU requesting consultations with Brazil under the Dispute Settlement Understanding in June 2005 (Brown and Trachtman, 2009).

Before both a WTO Dispute Settlement Panel and the Appellate Body, Brazil argued that its measures were necessary to prevent the risks to human, animal and plant life and health arising from the accumulation of waste tyres. The risks cited by Brazil were twofold. Firstly, it argued waste tyres in disposal facilities or abandoned across the countryside increase exposure of communities to toxic emissions caused by tyre fires. Secondly, tyres stored or abandoned out in the open are ideal breeding cited for mosquitoes, and thus facilitate the transmission of mosquito-borne diseases, such as dengue, yellow fever and malaria (Gray, 2008).

On 12 June 2007, the WTO Panel in Brazil-Measures Affecting Imports of Retreaded Tyres (henceforth 'Panel') found that the Brazilian ban on the importation of retreaded tyres from non-Mercosur countries violated the General Agreement on Tariffs and Trade (GATT) 1994. The Panel determined that the ban could not be justified under GATT Article XX which lays out specific instances in which WTO members may be exempted from GATT rules (Van Damme, 2008).

The Panel's interpretation of Article XX was subsequently appealed by the EC, despite the initial ruling having supported the EU's position. It argued that the Panel should have examined whether the ban actually contributed to the protection of human, animal or plant life or health, and if, therefore, the ban was 'necessary'. The Appellate Body reinforced the Panel's findings but was vague on the concept of necessity, opening up the possibility of future disputes around the wording of Article XX (Van Damme, 2008).

Nevertheless, the Panel and Appellate Body judgements meant that as a WTO Member, Brazil could not ban the importation of retreaded tyres from the EU while permitting them from Mercosur countries.

8.2.2. *Traditional gains from trade*

To determine the traditional gains from trade, GT (recalling Equation (7.3)) requires the parameters outlined in Section 7.2. These parameters and the data used to represent them in the model are detailed in Table 8.1.

Table 8.1. Model parameters — Brazil-retreaded tyre imports.

Description with Reference to Figure 7.1	Value
Elasticity of supply[a]	Pert(2.3,2.5,2.7)
Elasticity of demand[a]	Pert(–1.7,–1.5,–1.3)
Closed-economy quantity supplied, q_0 (#)[b]	6,750,000
Closed-economy, or autarchic price, p_0 ($/tyre)[b]	9.00
Free trade price, p_2 ($/tyre)[c]	3.00
Post-quarantine, or import price, p_1 ($/tyre)[b]	3.00
Wholesale marketing margin[d]	Pert(10%, 20%, 30%)
Retail marketing margin[d]	Pert(30%, 45%, 60%)

[a]Specified with reference to Galarraga and Markandya (2003).
[b]WTO (2007).
[c]Assumes negligible quarantine treatments are applied to imported tyres such that $p_1 = p_2$.
[d]Arbitrary values.

Table 8.2. The gains from trade to Brazil from imported tyre retreads.

	Mean	Standard Deviation
Change in consumer surplus (ΔCS)	\$105,705,000	\$5,604,000
Change in producer surplus (ΔPS)	–\$6,750,000	\$2,285,000
Gains from trade (GT)	\$98,956,000	\$6,355,000

8.2.3. *Results and discussion*

8.2.3.1. *Welfare analysis: Benefits from trade*

The net gains from trade resulting from the import of retreaded tyres to Brazil are listed in Table 8.2.

With the demand and supply curves used in this analysis, the results indicate that the net gains from trade to the Brazilian economy from allowing EU-retreaded tyres into the country are approximately \$99.0 million per year with a standard deviation of \$6.4 million.

8.2.3.2. *Indicative ecosystem services benefits*

Ecosystem services valuation is a tricky business, particularly in the case of policy evaluation. Ostensibly, it involves four steps (Freeman III, 2002; Plummer, 2009): (i) determine the how a policy will alter ecosystem structure and function; (ii) determine what impact this will have on the delivery of ecosystem services; (iii) place a value on a marginal unit of ecosystem services change; and (iv) aggregate the marginal change over the affected sites.

Each of these steps present an array of problems. In the context of interlinked ecological, agricultural and urban systems that might be affected by changing volumes of waste tyres, prediction is a risky business. Completing Steps (i) and (ii) in the face of uncertain and nonlinear subsystem interactions, predicting how external factors like waste tyre volumes might affect the collective system is a feat in itself. This only brings us to Step (iii), at which point we could probably be forgiven for taking our bat and ball and going home.

On the way, we might also be forgiven for stopping in at an increasingly popular ideological watering hole; a quaint little establishment known as the *Benefit Transfer*. This is an abode where many a tired, frustrated environmental economist has sought solace. The products on offer there avoid the complicated research needed to complete Step (iii) rigorously (whatever 'rigorously' means with complex systems), and instead provide a stopgap. It is a place that does a roaring trade when pressure is high, time short and decisions impending.

Simply put, benefit transfer involves taking ecological value information captured at one time and place and using it to estimate the value of environmental goods and services at another time and place (Wilson and Hoehn, 2006). This implies a degree of homogeneity between ecosystems that can lead to serious errors (Ready *et al.*, 2004; Rosenberger and Stanley, 2006; Kaul *et al.*, 2013; Kirchhoff *et al.*, 1997), although adjustments can be made for demographic and biophysical differences between sites (Ready and Navrud, 2006). It also infers the value put forward in the original analysis is accurate, but since we have no idea what the 'true' value is all we can hope for is that the methodology used is sound.

While acknowledging benefit transfer is controversial, we simply highlight here its potential evidentiary use in trade negotiations involving ecosystem services changes. Benefit transfer does allow economists to put values on ecosystem services in their reports for deliberation, instead of having to use the time-honoured cop out 'ecosystem impacts are beyond the scope of this analysis'.

Essentially, there are two different ways of forming these values presented in the literature. One advocates the use of avoided costs to estimate ecosystem services values and the other recommends using willingness-to-pay to prevent ecosystem services damage. We will not provide a critique of these approaches here and do not subscribe to one or other, but we will use the traditional avoided cost method for illustration in our case study of retreaded tyres.

With an avoided cost approach, an estimate of the marginal ecosystem services change resulting from the Brazilian ban relies on estimating the damages prevented by the policy. Now, proving cause and effect requires a good deal of guesswork, but to reemphasise, we are simply demonstrating how the technique potentially combines with our trade model to enable an overall evaluation of social welfare impacts.

For now, we shall leave to one side risks to animal and plant life and health pointed out by Brazil when banning imports of retreated tyres and concentrate on risks to human health. Specifically, let us look at the number of deaths from malaria, dengue fever and yellow fever. Brazilian deaths attributable to malaria in 2009 were 79 (World Health Organization, 2010b). Dengue fever deaths averaged 160 per year between 2009 and 2014 (BBC News, 2015). And, yellow fever deaths were 154 between 2017 and 2018 (World Health Organization, 2018). Based on these figures and ignoring likely increases in deaths as the diseases spread, assume approximately 400 people die of mosquito-borne diseases in Brazil each year. Furthermore, let us assume that one quarter (i.e. 100) of these deaths would have been avoided were it not for retreaded tyre imports.

In addition, we use the economic concept of *value of a statistical life* (VSL). The VSL is a measure of the willingness of society to pay for a reduction in mortality risk sufficient to lower the expected number of fatalities by one over a given period of time. It therefore represents the marginal rate of substitution between wealth and mortality risk for that period (Viscusi, 1993). The implication that money and life are substitutes at all is morally questionable, and consequently only a handful of VSL estimates have been put forward for a small number of countries. The authors are not aware of any such estimates for Brazil.

VSLs are available for Australia, however, so we will perform a simple conversion to estimate the VSL for Brazil.[1] Estimates relevant to Australia vary between $3 million and $15 million, but a review of these studies in 2007 concluded a value of $3.5 million was most appropriate (Abelson, 2007). In 2014, this value was revised to $4.2 million to correct for inflation, equivalent to US$3.0 million (Office of Best Practice Regulation, 2014). We can then convert this value to a Brazilian equivalent by using the Gross Domestic Product (GDP) per capita between the two countries. World Bank (2017) lists GDP per capita for Brazil as 15,500 international dollars and Australia as 47,000 international dollars, making the ratio between them approximately 33%.

Hence, we estimate the VSL for Brazil to be US$1.0 million, which puts our estimate of avoided costs of mosquito-borne disease from the retreaded tyre ban at US$100 million per annum. However, this ignores the avoided cost of non-fatal cases, as well as the avoided cost of preventative actions taken by the Brazilian government to try to reduce the extent of infection. For the purpose of this illustration, let us assume that these additional costs have the effect of doubling the avoided cost resulting from the retreaded tyre ban to US$200 million per annum. This provides an indicative cost of mosquito-related illness and death avoided by Brazil banning imports of

[1]Other countries for which VSL estimates are available include Austria, Canada, Denmark, France, Japan, New Zealand, South Korea, Sweden, Switzerland, Taiwan, UK and the USA (Miller, 2000).

retreaded tyres, although as noted above, it does not include environmental costs related to toxic emissions.

8.2.3.3. *Expected net change in social welfare*

Plotting avoided ecosystem services damage with the gains from trade (or lack thereof), as we did in Chapter 6, provides a visual representation of the net change in social welfare anticipated from banning retreaded tyre imports to Brazil. In Figure 8.1, the mean net change in social welfare represented by changes in consumer and producer surpluses is shown over a 30-year period with the expected ecosystem services cost avoidance and gains from trade. Values are expressed in US dollars and future values are discounted at 8% per annum.

Table 8.3 summarises gains from trade, avoided ecosystem services damage and net social welfare change over 10-, 20- and 30-year-time periods. It can be seen from this table that with a gain in social welfare occurring because of the retreaded tyre ban over the full course of 30 years, the period being considered is not critical. Moreover, the cumulative net change in social welfare from banning imports increases from $778.7 million over 10 years to $1238.0 million over 30 years.

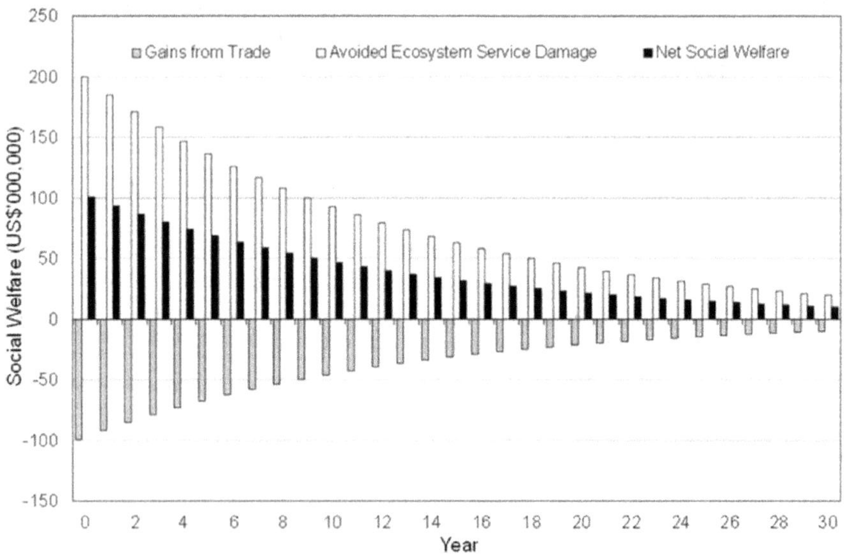

Figure 8.1. Change in social welfare over time resulting from Brazilian-retreaded tyre imports.

Table 8.3. Cumulative net change in social welfare from Brazilian ban on imported tyre retreads.

Year	Gains from Trade — Brazil ($ Million)	Avoided Ecosystem Services Damage ($ Million)	Net Change in Social Welfare ($ Million)
10	–763.3	1,542.0	778.7
20	–972.0	1,963.6	991.6
30	–1,213.5	2,451.6	1, 238.0

These results suggest that the WTO ruling on the Brazilian-retreaded tyre import ban, whereby the ban was deemed to be in breach of Article XX of the GATT, led to an undesirable outcome in terms of the overall impact on social welfare. Overturning the ban meant that the importing country was exposed to human health costs that exceeded the gains from trade by over US$1.2 billion over a 30-year period. Indeed, the estimated loss in ecosystem services would need to more than halve for the decision to overturn the ban to breakeven in terms of net social welfare. Alternatively, in including the gains from trade of the exporting countries, the total gains from trade would need to more than double.

8.3. Measures Affecting Tuna Imported from Mexico

8.3.1. *Revisiting the issue*

In this section, we discuss the issue presented in Chapter 4 regarding measures placed on tuna imported to the US from Mexico. We learned from that chapter that this is a particularly interesting case because it involves measures placed on a product due to the perceived negative environmental impacts of its production method.

The method in question is known as purse-seine fishing, which involves a net called a seine deployed from a fishing trawler by a smaller boat known as a seine skiff (or panga boat) to encircle schools of fish. In the case of yellowfin tuna, when a school is located the skiff deploys the seine around the perimeter. The net is weighted on one end and buoyed on the other, so it hangs vertically in the water. It is then sealed at the bottom by tensioning a line arranged like a drawstring on a coin purse (hence the name 'purse-seine fishing'). The net is then winched in by the trawler and the ensnared fish are deposited on the deck for sorting.

The concern with purse-seine fishing raised by the US in this particular dispute relates to bycatch. In addition to tuna, any other marine creatures that are not small enough to escape the weave of the net are also caught. These include other commercial species of fish, but also species of sharks and rays, sea turtles, marlins and sailfish, and on occasion dolphins (Hall and Roman, 2013). Dolphins caught in the sein are often drowned before they can be released, causing significant declines in their population and reproductive rates (Cramer *et al.*, 2008). The highest losses were reported between 1959 and 1972, when it is estimated purse-seine fishing caused the deaths of approximately 4.9-million dolphins, an average of almost 350,000 per year (Wade, 1995).

A proportion of the tuna consumed in the US originates from an area lying off the coasts of California, Mexico, Central America, Colombia, Ecuador, Peru and Chile known as the Eastern Tropical Pacific. Here, dolphin bycatch is a particular issue as schools of yellowfin tuna (*Thunnus albacares*) often swim beneath pods of pantropical spotted dolphins (*Stenella attenuate*) that in turn are circled by seabirds (Scott *et al.*, 2012). While the precise reason for such behaviour in this particular region is unclear, fishers have long exploited these associations as seabirds and dolphins are far easier to observe than tuna swimming at depth beneath them. When dolphins are detected, the seine is used to encircle both the pod and the yellowfin tuna. A manoeuvre called the 'backdown' is then performed by the fishers to release the trapped dolphins.

The backdown consists of putting the trawler in reverse and pulling the net taut such that it elongates and then sinks several metres below the surface in the farthest section from the boat. This allows dolphins to exit the net, while at the same time the vessel pulls the net under the pod (Francis *et al.*, 1992; Hall and Roman, 2013). However, although backdowns succeed in releasing many trapped animals, dolphin fatalities still occur. Factors such as the time of day of a catch greatly influence the effectiveness of the manoeuvre with sundown catches producing higher mortality rates than day catches (Francis *et al.*, 1992). Inter-American Tropical Tuna Commission (1989) reported that of almost 16,000 observed catches 9.3% occurred near sundown, yet these accounted for 26% of recorded dolphin fatalities (Francis *et al.*, 1992).

Following concern expressed by biologists, the US Congress enacted the Marine Mammal Protection Act in 1972, which addressed the dolphin bycatch issue in the Eastern Tropical Pacific (Brown, 2005). According to this Act, any country wishing to export tuna to the US must first demonstrate that it meets the same dolphin protection standards that must be met by American fishers. If it does not, the Act requires that an embargo be placed on all imports of the fish from that country and any intermediary countries handling the tuna for processing or canning.

The Marine Mammal Protection Act addressed the effects of commercial fishing by US vessels on ocean mammals in general and, following several amendments in the 1980s, the effects of yellowfin tuna fishing on dolphins specifically (Parker, 1999). It imposed a moratorium on the hunting, capturing or killing of marine mammals except in cases of incidental harm in the course of commercial fishing operations conducted under a permit issued by the National Marine Fisheries Service (Musgrave and Stephens, 1993). Only one permit has been issued to the American Tunaboat Association allowing up to 20,500 dolphins to be taken annually (Musgrave and Stephens, 1993).

Despite attempts by Mexico to comply with these requirements, the US placed an embargo on yellowfin tuna and tuna products from Mexico in 1990 (McDorman, 1992). In November 1990, Mexico, through the GATT, sought consultations with the US about the trade aspects of the embargo. When these were unsuccessful, Mexico requested that a GATT Panel of experts be asked to look into the dispute.

A three-member Panel was established in February 1991. In evidence put before them, Mexico contended that the US embargo contradicted the ban on quantitative prohibitions under Article XI of GATT, the national treatment provision under Article III, and the exceptions granted under Article XX (Joyner and Tyler, 2000). The US maintained that the ban it had imposed was in the interests of marine conservation and should therefore be accepted.

Complicating matters, the US Congress passed the Dolphin Protection Consumer Information Act (1990) that made it a violation of the Federal Trade Commission Act to use any label suggesting tuna had been caught in a manner not harmful to dolphins if the tuna was harvested by either driftnets or purse-seines (Parker, 1999; Brown, 2005). In February 1991, just as the Panel had been established to look into the first embargo, the environmental group Earth Island Institute won a court-ordered second embargo against yellowfin tuna from Mexico displaying a 'dolphin-safe' label on the grounds that this was a violation of the Federal Trade Commission Act (Parker, 1999). The second embargo took effect in March 1991, and a further court-ordered embargo was imposed on 13 intermediary nations that received tuna from Mexico for packaging in January 1992 (Parker, 1999). These included Costa Rica, Italy, Japan, Spain, France, the Netherlands, the United Kingdom, Canada, Colombia, the Republic of Korea, and members of the Association of Southeast Asian Nations.

The Panel handed down its report in September 1991. While acknowledging it had been designed to conserve dolphins, the Panel determined that the US embargo on Mexican tuna was inconsistent with the GATT. Famously, its verdict stated that no country could restrict imports in any manner for the purpose of protecting the environment outside its own jurisdiction (GATT, 1991).

The US was understandably dissatisfied with the Panel's ruling on measures put in place for environmental protection, but because the Panel upheld the embargo on intermediary countries, Mexico too was also unhappy with the dispute resolution. This catalysed negotiation of a multilateral agreement on the protection of dolphins that would be mutually beneficial to all countries operating within the Eastern Tropical Pacific tuna fishery. The La Jolla Agreement for the Reduction of Dolphin Mortality in the Eastern Pacific Ocean was subsequently formed in 1992 and endorsed by the 1995 Declaration of Panama signed by 12 nations participating in the Eastern Tropical Pacific tuna fishery (Brown, 2005). These included Belize, Columbia, Costa Rica, Ecuador, France, Honduras, Mexico, Panama, Spain, the US, Vanuatu and Venezuela (Brown, 2005).

In addition, the International Dolphin Conservation Programme was founded at the suggestion of the Inter-American Tropical Tuna Commission, consisting of a series of quotas on dolphin fatalities and a stringent monitoring programme. The Program was written into American law through the adoption of the 1992 International Dolphin Conservation Programme Act (Brown, 2005). This amendment to the US Marine Mammal Protection Act relaxed the definition of 'dolphin-safe' and enabled tuna from foreign producers to be imported into the US provided that it was harvested according to the rules and guidelines of the International Dolphin Conservation Programme (Brown, 2005).

8.3.2. *Traditional gains from trade*

The parameters needed to determine the traditional gains from trade, GT, (i.e. Equation (7.3), Section 7.2), are listed in Table 8.4.

8.3.3. *Results and discussion*

8.3.3.1. *Welfare analysis: Benefits from trade*

The net gains from trade resulting from the import of Mexican tuna to the US are listed in Table 8.5. This represents a movement from a closed-economy situation under the US embargo to one of quarantine-restricted trade whereby a tariff remains in place.

Table 8.4. Model parameters — US tuna imports.

Description with Reference to Figure 7.1	Value
Elasticity of supply[a]	Pert(0.1,0.2,0.3)
Elasticity of demand[a]	Pert(–0.7,–0.5,–0.3)
Closed-economy quantity supplied, q_0 (T)[b]	260,000
Closed-economy, or autarchic price, p_0 (US$/T)[c]	913
Free trade price, p_2 (US$/T)[d]	647
Post-quarantine, or import price, p_1 (US$/T)[c]	809
Wholesale marketing margin[e]	Pert(5%,10%,15%)
Retail marketing margin[e]	Pert(10%,15%,20%)

[a]Cheng and Capps Jr. (1988), Yen and Huang (1996) and Babula and Corey (2004).
[b]Total demand for canned tuna in the US in 2014 is reported as approximately 330,000 tonnes in WTO (2016), which is equivalent to 1.03 kg per capita. As a rough estimate, assuming that per capita tuna consumption to have remained constant over time (although it probably has not due to changing incomes, preferences and availability of substitutes), the total demand for canned tuna in the US in 1991 is estimated at 260,000 tonnes.
[c]The US banned Mexican yellowfin tuna imports from 1980 to 1986, in retaliation for the seizure of US tuna boats fishing within Mexico's 200-mile coastline. Over this period, the tuna price averaged US$913/tonne. When the ban was lifted in 1986, the price fell to $809/tonne (Sakagawa, 1991). This included a 12.5–35% tariff (Murphy, 2006).
[d]Assumed to be 25% below the post-quarantine price. Since 1972, the US tariff on imported tuna had been 12.5% ad valorem on canned tuna packed in water and 35% on canned tuna packed in oil (Murphy, 2006).
[e]Approximated using later estimates from WTO (2016).

Table 8.5. The gains from trade to the US from imported canned tuna from Mexico.

	Mean	Standard Deviation
Change in consumer surplus (ΔCS)	$42,729	$1,649,000
Change in producer surplus (ΔPS)	–$28,388,000	$1,443,000
Gains from trade (GT)	$71,117,000	$2,722,000

With the demand and supply curves used in this analysis, the results indicate that the net gains from trade to the US economy from allowing canned tuna from Mexico into the country are approximately $71.1 million per year with a standard deviation of $2.7 million.

8.3.3.2. *Indicative ecosystem services benefits*

Recall from Section 8.2.3.2 that there are two different ways of forming ecosystem services values: avoided costs and willingness-to-pay. Once again, we do not advocate one approach over the other and are only interested in providing a broad, indicative estimate of ecosystem services values for the purposes of illustration. Since we used the avoided cost methodology to provide an estimate of the value of ecosystem services impacts in the Brazil-imported tyres case study, we will do so again in this example to be consistent.

Haraden *et al.* (2004) present shadow prices for dolphins in the Eastern Tropical Pacific fishery that reflect both the private and social ecosystem services costs incurred when they are killed in the process of fishing. Private provisional ecosystem services costs are incurred by tuna fishers as dolphins indicate the location of yellowfin tuna shoals, while social ecosystem services costs include the loss of cultural and recreational benefits of dolphins in the Eastern Tropical Pacific. Between 1985 and 1987, Haraden *et al.* (2004) estimated that preventing the death of a single dolphin avoids average costs of US$1,986.

In forming a broad estimate of ecosystem services changes, we assume that 15% of Mexican tuna harvested in the Eastern Tropical Pacific fishery was supplied to the US in the early 1990s (Murphy, 2006). Further assume that 80,000 dolphin mortalities per year were attributable to non-US fishers around this period (Sakagawa, 1991). This implies that approximately 12,000 dolphin mortalities occurred as a direct result of tuna harvested for the US market. Assuming constant marginal costs, using the Haraden *et al.* (2004) shadow price the estimated total ecosystem services damage prevented by the US embargo is $23.8 million per year.

8.3.3.3. *Expected net change in social welfare*

As in the Brazilian-retreaded tyre case, plotting avoided ecosystem services damage with the gains from trade allows us to clearly see the net change in social welfare anticipated from the US tuna embargo. In Figure 8.2, the mean net change in social welfare represented by changes in consumer and producer surpluses is shown over a 30-year period with the expected ecosystem services cost avoidance and gains from trade. Values are expressed in US dollars and future values are discounted at 8% per annum.

It is doubtful an embargo would last as long as 30 years, but we show results over this period to be consistent with the previous case studies. Table 8.6 summarises gains from trade, avoided ecosystem services damage and net social welfare change over 10-, 20- and 30-year time periods. The cumulative net reduction in social welfare increases from –$364.6 million over 10 years to –$579.6 million over 30 years.

These results suggest that the Panel's ruling on the US embargo on Mexican led to a net social welfare gain. By determining the embargo

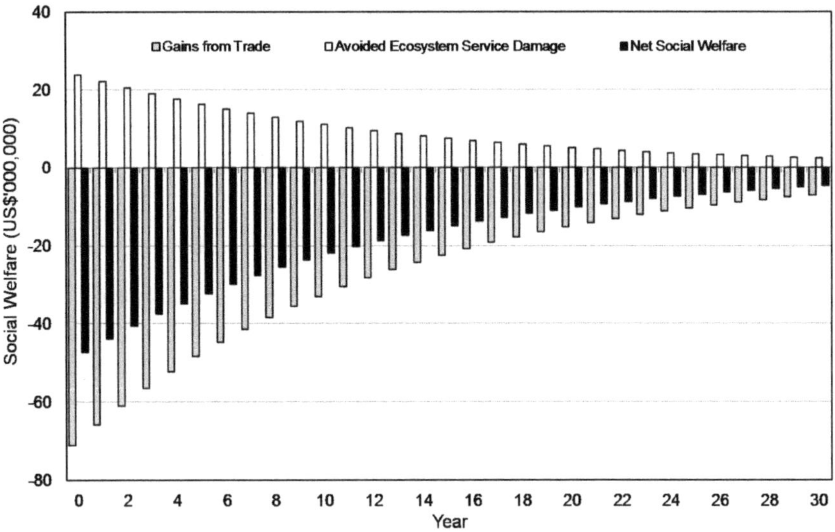

Figure 8.2. Change in social welfare over time resulting from an embargo on Mexican tuna imported to the US.

Table 8.6. Cumulative net change in social welfare from US embargo on Mexican tuna.

Year	Gains from Trade — US ($ million)	Avoided Ecosystem Services Damage ($ Million)	Net Change in Social Welfare ($ Million)
10	–548.3	183.7	–364.6
20	–698.2	234.0	–464.3
30	–871.7	292.1	–579.6

to be inconsistent with the GATT, the net gains from trade for the US are substantially larger than the ecosystem services costs associated with increased numbers of dolphin deaths. So much so that estimated ecosystem services costs would need to increase by a factor of three just to breakeven in terms of net social welfare. Moreover, if the gains from trade for the exporting country are included, then this break-even factor would need to be even larger.

The situation depicted in Figure 8.2 and Table 8.6 contrasts with that depicted in Figure 8.1 and corresponding Table 8.3. In the dolphin–tuna case, the ecosystem benefits of the trade embargo on Mexican tuna are insufficient to offset the loss in traditional gains from trade, while the opposite is true in the retreaded tyre case in which the traditional gains from trade lost as a result of the Brazilian trade ban were less than the ecosystem services gains. It follows that our indicative results suggest that the WTO ruling in the dolphin–tuna case was indeed justified on social welfare grounds, but its ruling on the retreaded tyre case led to considerable social welfare losses to Brazil as the importing country.

8.4. Conclusion

This chapter has applied the methodology developed in Chapter 7 which introduces ecosystem services changes into trade analysis to give a more complete social welfare evaluation of the impacts of trade. Applying this methodology in the context of the Brazilian ban on retreaded tyres and the US ban on Mexican tuna outlined in Chapter 4, we have demonstrated in broad terms how such a methodology might potentially be used in WTO dispute settlement to indicate which trade policies are likely to lead to the best social welfare options, rather than simply the best producer outcomes.

In both of these examples, however, we wish to reemphasise that the estimates of ecosystem services impacts provided to generate the results are only intended as indicative numbers to illustrate application of the methodology developed in Chapter 7. They do not represent the results of an exhaustive search of the literature and detailed ecosystem services evaluation. In practice, we would hope to see more careful studies comparing ecosystem services and producer and consumer welfare changes put before the WTO to enable it to determine which trade policies lead to the most desirable social welfare outcomes.

However, while the quantification of these case studies has been primarily illustrative, this chapter has demonstrated in broad terms how the methodology of Chapter 7 can lead to contrasting outcomes from historical WTO trade dispute rulings. In particular, while for both case studies the actual WTO ruling was to support trade, the analysis in this chapter has suggested that in the Brazilian-retreaded tyre case study, the estimated ecosystem services damages prevented from banning imported retreaded tyres were larger than the traditional gains from trade forfeited, implying that allowing trade results in a net loss of social welfare and thereby contradicting the WTO ruling. Whereas in the case of the US/Mexican tuna dispute the estimated prevented ecosystem services damages from banning Mexican tuna are smaller than the traditional gains from trade lost as a result of the ban, implying that allowing trade results in a net gain in social welfare and thereby supporting the WTO ruling.

Given we have added complexity to the assessment though, what might this mean in practical terms? Have we merely introduced another impediment to free trade by allowing trade-restrictive countries more leverage and legal ammunition? After all, in many cases, ecosystem services are unpriced and assigning values to them can be difficult, even with simplistic methods like benefit transfer. So, how could it all work in practice if the WTO were to really include ecosystem services assessments into trade negotiations and dispute settlement?

We will tackle these issues in Chapter 9, drawing from experiences with cost share agreements for biosecurity responses in Australia.

Part D

A Way Forward

Chapter 9

Dealing with Complexity: Non-Monetised Impacts and Distributional Considerations

9.1. Introduction

The US imposition of steel tariffs in early 2018 set off a chain of events that made many economists shift uncomfortably in their seats. As the world's largest exporter of steel (World Steel Association, 2018), China stood to be worst affected, and as a significant and growing market for many US-manufactured goods, she could be expected to retaliate with tariffs of her own. By year's end, the US had introduced tariffs on US$250 billion of Chinese goods and China had indeed responded with duties on US$110 billion worth of US goods (Bryan, 2018). Coming as it does in the wake of the 2008 stock market crash, this new protectionism is eerily similar to policies that exacerbated the Great Depression following the 1929 stock market crash, precipitating the Second World War.

The intriguing G7 summit of June 2018 highlighted the fragility of international relations with respect to trade. The summit failed to resolve disagreements stemming from the imposition of US tariffs. Instead, it produced arguments between traditional allies Canada and the US over Canadian tariffs on agricultural goods and retaliatory measures, fanning protectionist flames rather than mollifying them. It is interesting to note that although tariffs dominated discussions, G7 countries mete out large subsidies to their agricultural producers that also distort markets (OECD, 2018).

As unsettling as these recent protectionist trajectories are, the situation need not necessarily be all bad news. Indeed, from the point of view of ecosystem services, protectionism that leads to less ecosystem services impacts from production and transport could actually be a good thing. After all, these impacts are rarely, if ever, reflected in prices. If they were, the benefits of world trade would be substantially less than is reported. Considering the US withdrawal from the 2015 Paris deal to combat climate change, perhaps tariff increases that lower the ecosystem services impacts of world trade will give something back to the environment.

The problem is that ecosystem services impact reductions achieved in this way will have been a positive externality of the process, rather than a designed outcome, and as such will be under-produced. In this chapter, we discuss ways in which the World Trade Organization (WTO) might explicitly consider ecosystem services impacts when ruling on trade disputes so as to internalise ecosystem services externalities. More specifically, in Section 9.2, we draw inspiration from biosecurity policies in Australia that seek to share the costs of eradicating pest and disease outbreaks between public and private beneficiaries. We then propose in Section 9.3 an ecosystem services penalty system the WTO could use to regulate multilateral ecosystem services impacts from trade. In Section 9.4, we outline the benefits of introducing a consensus-building feature into this proposal whereby the WTO uses Multi-Criteria Decision Analysis (MCDA) to determine if the size of multilateral ecosystem services impacts related to a good's production and transport to market outweigh traditional gains from trade. This is followed in Section 9.5 by a detailed elaboration of how the MCDA approach would operate in this context. This chapter ends with a brief summary and conclusion.

9.2. Import Risk Analyses and Cost Sharing Agreements in Biosecurity

If it is to embed ecosystem services impacts within trade negotiations, the WTO must have a way of judging their size next to producer and consumer surplus changes. In Section 7.3 of Chapter 7, we outlined the six possible results of such a comparison and what this would mean in terms of a trade decision, but we stopped short of saying exactly how this could be done in the absence of ecosystem services values impacted by trade. Traditional gains from trade are relatively straightforward to value, but we saw in the case studies of Chapter 8 that ecosystem services tend both to be more complex and to feature considerable uncertainty in terms of value.

So, how might a Dispute Settlement Panel determine the extent of ecosystem services changes alongside traditional gains from trade? Well, to a certain extent, they already do. As seen in Chapters 3, 5 and 6, in assessing the legitimacy of Sanitary and Phytosanitary Standard (SPS) measures, Import Risk Assessment (IRA) underpinning those measures is taken into consideration. An IRA essentially assesses the probability of invasive species entry, establishment and spread as a result of trade, and the consequences should they do so. For example, in Australia, The *Biosecurity Act 2015* currently (and the *Quarantine Act 1908* previously) requires Australia's Director of Animal and Plant Quarantine to ensure environmental factors are considered in the consequence assessment process (Department of Agriculture and Water Resources, 2016; Australian Government Department of Agriculture Fisheries and Forestry, 2011).

The method of consequence assessment used is relatively simplistic and contains ambiguities. All economic, environmental and social consequences are grouped together and determined to be 'negligible', 'very low', 'low', 'moderate', 'high' or 'extreme' (Department of Agriculture and Water Resources, 2016). Linguistically, there is uncertainty as to what each of these classifications means. It is also unclear as to what timeframe should be used to assess consequences. While economic consequences might be immediate, environmental consequences may take hundreds of years or more to manifest.

Despite its imprecision, the point to make here is that a process is routinely used in a trade policy context that deals with uncertain consequences of trade, albeit only in terms of an importing county's potential invasive species costs.

We can also look to Australia for a second example, this time related to funding incursion responses when there are a mix of public and private beneficiaries. When the circumstances surrounding the detection of an invasive species are such that eradication is technically and economically feasible, Australia has unique cost sharing arrangements in place. The Emergency Animal Disease Response Agreement (EADRA) and Emergency Plant Pest Response Deed (EPPRD) involve invasive organisms being placed in one of four cost sharing categories relating to their potential impact on public resources and private industries (AHA, 2002; PHA, 2005).

If a species categorised under one of these agreements is detected, the category chosen dictates an appropriate allocation of property rights over (potentially) affected public and private capital. Eradication funding obligations for government and private industry are then specified according to this allocation. A Category-1 species classification means that 100% of the benefit of eradication accrues to the public, and as such, all costs are to be paid by the government. Categories 2–4 species affect smaller proportions of public assets compared to private, and as such, government contributions to eradication fall to 80%, 50% and 20%, respectively.

Provided the categorisation process is objective and collection mechanisms are in place to obtain funds from beneficiaries, there is no danger of parties free-riding on biosecurity services provided by others. Importantly, sharing the burden of response also means there is sufficient capacity to provide appropriate eradication funding according to the circumstances.

However, this model too has its drawbacks. For one thing, it is rarely used because eradication is seldom technically or economically feasible. For another, recovering private beneficiary contributions to eradication can involve high-transaction costs, particularly in the case of polyphagous species that affect a large number and diversity of industries, each with differing capacities to pay. Nevertheless, eradication cost sharing agreements are another example of policies designed around systems that are highly complex.

9.3. An Ecosystem Services Penalty System

Using the above examples as inspiration, a categorisation approach might present a possible policy centrepiece for WTO Dispute Settlement Panels of the future. In the same way that the consequences of invasive species to an importing country are assessed as part of an IRA, it may be possible to use a similar assessment of net ecosystem services changes likely to result from trade in both the importing and exporting countries. Likewise, just as the full cost of eradication is not known and is expressed as public and private proportions when the EADRA or EPPRD are invoked, ecosystem services impacts related to trade might also be expressed as a proportion of easily measured traditional gains from trade.

In Chapter 7, we speculated as to what a WTO Dispute Resolution Panel might consider if it was to adopt a multilateral view of welfare change (W) encompassing both multilateral gains from trade (GT_{M}) and multilateral ecosystem services impacts (ES_{M}). We established that W is the sum of GT_{M} and ES_{M} (i.e. Equation (7.11)), and that, ideally, a Panel's decision will result in a net social welfare gain.

We went on to outline some of the difficulties involved in valuing ecosystem services in Chapter 8, or more particularly changes in them resulting from possible invasive species introductions. Section 8.2.3.2 in particular showed there is a serious ES_{M} information gap the Panel must overcome.

Without monetised information concerning the size (and sign) of ES_{M}, the full social welfare costs of trade to the communities of two trading partners cannot be ascertained. However, in contrast, GT_{M} is readily quantifiable as the sum of gains from trade in both the importing and exporting countries and might be used as a benchmark to compare uncertain ES_{M} when deciding whether or not to allow trade.

Given this, consider the Dispute Settlement Panel's role as one in which the appropriate size of an ecosystem services impact penalty must be determined on the basis of qualitative information. This penalty could take on a semi-quantitative form in the same way Australia's IRAs categorise consequences of invasive species, and the EADRD and EPPRD categorise the mix between public and private beneficiaries.

Suppose the penalty, α, is to be applied to the absolute value of GT_{M} up to a point the Panel feels represents ES_{M}:

$$ES_{\mathrm{M}} = \alpha|GT_{\mathrm{M}}| \tag{9.1}$$

Here, α is an ecosystem services penalty that takes on a value relative to the severity of ecosystem services impacts resulting from trade but represented as a proportion of the absolute value of the multilateral gains from trade, $|GT_{\mathrm{M}}|$. If α is positive, it implies net ecosystem services costs will result from trade, and if α is negative, it implies net ecosystem services benefits will result.

Under this penalty decision model, resolving on the sign and magnitude of α becomes a key part of the dispute resolution process, and we will discuss how this might be accommodated in Section 9.3. The welfare resulting from trade (i.e. Equation (7.11)) under this model is

$$W = GT_{\mathrm{M}} - \alpha|GT_{\mathrm{M}}| \tag{9.2}$$

The term $\alpha|GT_{\mathrm{M}}|$ is essentially a substitute monetary expression of ES_{M} that uses GT_{M} as a scaling device. The choice of penalty is drawn from a categorisation of different levels of severity of ecosystem services impacts and captures a myriad of complex information about ecological, agricultural and social systems potentially impacted by trade in both the importing and exporting countries. However, although the α term itself is an extreme simplification, the decision of what choice from the severity categorisation system should apply in different contexts can potentially involve large amounts of information.

We suggest that this complexity can be unpacked through the use of a MCDA when deciding on an ecosystem services penalty rate. This method of analysis is designed to simplify complex policy problems involving multiple stakeholders, several possible outcomes and a range of incompatible criteria by which to assess the outcomes (Proctor and Drechsler, 2006). We argue below that it may be possible to enhance the current use of expert testimony, scientific literature and producer risk documentation by Dispute Settlement Panels through the use of MCDA.

9.4. Outlining the Benefits of Using Multi-Criteria Decision Analysis

Given the complex mixture of market and non-market trade impacts, MCDA techniques may offer a practical solution to trade policy-makers by stimulating discussion about possible policy choices, trade-offs and uncertainties. MCDA identifies solutions to complex and uncertain decisions characterised by alternatives/options that can be evaluated using weighable criteria (Jankowski and Nyerges, 2001). In the context of a decision of whether to allow trade in a particular commodity or not, the relative performance of each option (i.e. trade, no trade, limited trade) against the weighted criteria set allows policy-makers to identify the optimal alternatives. In the broader context of global trading, life-cycle assessments and green national accounts, two widely-used accounting tools, could be incorporated to increase trading nations' awareness of the use and the limits of ecosystems (Koellner *et al.*, 2011; Weber, 2011).

Terms such as *multi-criteria decision analysis*, *multi-attribute decision-making*, *multiple objective decision methods* and *multi-objective decision-making* are often used interchangeably to refer to a set of techniques that involve assessment of continuous alternatives and incorporate mathematical optimisation methods (Yoon and Hwang, 1995; Rietveld, 1980; Kazana *et al.*, 2000). We will use the term MCDA to refer to a suite of approaches designed for situations where more than one objective or criterion must be considered within a decision-making context, and where there are a finite number of pre-determined and discrete options from which to choose.

MCDA is based on the premise that decision-making can be improved by making the criteria explicit and ranking options according to how well it satisfies each of the criteria. Options can literally be any discrete set of alternatives we wish to choose between, be it management options, investment portfolios, capital items, or in this particular case the size of ecosystem services penalty appropriate for trade in a particular good. Distinguishing between options can involve selecting one superior option above all others or ranking options from least-preferred to most-preferred.

Although techniques vary, there are a number of key steps that a MCDA involves. The first usually involves identifying feasible options to choose between in the decision problem. The choice of options (or scenarios) and overall objectives can be developed by various sources including an expert or lay jury, expert advice, computer simulation models, and/or political processes (Proctor and Drechsler, 2006).

Next, the decision-makers must clarify the objective to be achieved in the MCDA process and identify criteria by which to judge the options. Weights can be applied to each of the criteria to reflect how important each criterion is to decision-makers in relation to the overall objective. The weights provide a transparent bias which can be manipulated according to the idiosyncrasies of the decision or site affected by a decision. For the weighting process to function effectively, decision-makers should have access to a wide range of information and decision support tools so that they may arrive at an appropriate set of weights through a process of consultation and deliberation (Burgman *et al.*, 2006). If all criteria are equally important no weights are necessary.

The final step is to examine how each option performs against the weighted criteria and form a hierarchy of preferences using an aggregation method. This might be as simple as weighted averages of highest or lowest option scores or involve more complicated processes that include pairwise comparisons and variance. The sensitivity of the hierarchy can also be tested with respect to the assigned weights and aggregation method. The whole process can also be repeated with fine tuning by the decision-makers if results are not definitive.

By the end of the MCDA process, the selected option represents the most satisfactory course of action given the information and criteria used. It will not necessarily be the best option in terms of each individual criteria, but rather it performs the best overall. Moreover, the option that emerges as most-preferred only represents the best option for a particular group of decision-makers at a particular point in time in a particular decision context. Repeating MCDAs periodically may be required to take account of changing circumstances and new information.

In Section 9.5, we will go through each of the MCDA steps and suggest a possible design that a WTO Dispute Settlement Panel might use to make social welfare-enhancing decisions about when to allow trade and when to restrict it. As outlined previously, the basic objective of this process is to determine an appropriate ecosystem services penalty to apply in specific cases of trade. However, in the process of doing so the MCDA will facilitate an open and transparent approach to group learning and adaptive policy-making in which all parties share and gain knowledge about the potential economic, social and environmental effects of trade.

9.5. Designing a Multi-Criteria Decision Analysis to Determine an Appropriate Ecosystem Services Penalty

9.5.1. *Choosing options and the objectives*

The choice of options and of the overall objective are closely related steps in any decision-making process. The objective can be as broad as necessary, but when it comes to the WTO the primary objective is to choose the option which delivers the highest multilateral social welfare in terms of allowing or preventing trade. Hence, a Dispute Settlement Panel must reach overall agreement on whether multilateral social welfare is maximised by allowing trade subject to a different restriction options, or to prevent it altogether. The options could reflect each of the preferred scenarios of the Panel if they have entered into the process with preconceived ideas about the ecosystem services penalty that they feel should be applied. However, they could just as easily be based on an amalgamation of ideas.

Options should be sufficient in number to represent a realistic and varied selection for the decision-makers but should not be too numerous to make the analysis cumbersome and impractical. Furthermore, options can be altered and adjusted as part of an iterative process in which Panel members negotiate suitable values the penalty could take on.

As an example, a polychotomous ecosystem services penalty might be used that takes on a value of 150%, 50%, 0%, –50% or –150% depending on the relative severity of ecosystem services impacts that could potentially result from trade. Bearing in mind Equation (9.1), note that selecting a value of 150% or 50% implies the ecosystem services impacts of trade are seen as negative, while selecting a value of –50% or –150% implies the ecosystem services impacts of trade are seen as positive. To keep things simple, let us assume that the descriptors ‘high cost’, ‘moderate cost’, 'low cost', 'moderate benefit' and 'high benefit' are used to describe a 150%, 50%, 0%, –50% or –150% penalty value, respectively.

$$\alpha = \left\{ \begin{array}{rl} 1.5 & \text{if } ES_{\mathrm{M}} \text{ is 'high cost'} \\ 0.5 & \text{if } ES_{\mathrm{M}} \text{ is 'moderate cost'} \\ 0 & \text{if } ES_{\mathrm{M}} \text{ is 'low cost'} \\ -0.5 & \text{if } ES_{\mathrm{M}} \text{ is 'moderate benefit'} \\ -1.5 & \text{if } ES_{\mathrm{M}} \text{ is 'high benefit'} \end{array} \right\} \tag{9.3}$$

In terms of the six cases, we outlined in Chapter 7 where combinations of GT_M, ES_M and $GT_M + ES_M$ yield different trade results, we can deduce what choice of ecosystem penalty might apply:

(1) **$GT_{\mathrm{M}} > 0$ and $ES_{\mathrm{M}} > 0$:** This case implies a 'moderate benefit' or 'high benefit' ecosystem services penalty (i.e. $\alpha = -0.5$ or $\alpha = -1.5$) as ecosystem services are positively impacted by trade and reinforce positive multilateral gains from trade.
(2) **$GT_{\mathrm{M}} < 0$ and $ES_{\mathrm{M}} < 0$:** In this case, the ecosystem services costs are assessed to be 'high cost' or 'moderate cost' (i.e. $\alpha = 1.5$ or $\alpha = 0.5$), and whatever penalty applies will reinforce the negative multilateral gains from trade.
(3) **$GT_{\mathrm{M}} > 0, ES_{\mathrm{M}} < 0$ and $GT_{\mathrm{M}} + ES_{\mathrm{M}} > 0$:** A 'moderate cost' positive ecosystem services penalty would apply in this case (i.e. $\alpha = 0.5$). Multilateral ecosystem services impacts are negative, but they are judged by the WTO not to exceed the multilateral gains from trade.
(4) **$GT_{\mathrm{M}} > 0, ES_{\mathrm{M}} < 0$ and $GT_{\mathrm{M}} + ES_{\mathrm{M}} < 0$:** In this case, a 'high cost' ecosystem services penalty would apply whereby $\alpha = 1.5$. This represents a particularly challenging scenario for the WTO in that multilateral ecosystem services costs resulting from trade are assessed to be demonstrably higher than the multilateral gains from trade.
(5) **$GT_{\mathrm{M}} < 0, ES_{\mathrm{M}} > 0$ and $GT_{\mathrm{M}} + ES_{\mathrm{M}} < 0$:** This case implies a 'low cost' or 'moderate benefit' ecosystem services penalty (i.e. $\alpha = 0$ or $\alpha = -0.5$). Multilateral ecosystem services impacts from trade suggest low costs or even moderate benefits, but not to the extent that they outweigh multilateral losses from trade.
(6) **$GT_{\mathrm{M}} < 0, ES_{\mathrm{M}} > 0$ and $GT_{\mathrm{M}} + ES_{\mathrm{M}} > 0$:** In this case, a 'high benefit' ecosystem services penalty would apply whereby $\alpha = -1.5$. The ecosystem services benefits resulting from trade are high and warrant trade going ahead despite multilateral losses in gains from trade. Like Case (4), this case represents a challenge for the WTO in that it must

assess the associated ecosystem services benefits to be sufficiently high to justify consumers and producers collectively being disadvantaged by trade going ahead.

In practice, using broad terms like 'high', 'moderate' and 'low' in descriptors can be problematic and we merely use these for simplicity. Ambiguous use of language can be particularly pervasive where the same term is interpreted differently by participants, resulting in misunderstanding and arbitrary disagreement (Carey and Burgman, 2008; Webb and Raffaelli, 2008). It is therefore important that Panel members deliberate over the descriptors used and clearly understand their meaning (Liu *et al.*, 2010b).

9.5.2. *Selecting criteria for choosing the penalty option*

Criteria should be constructed to compare each of the options and must, therefore, relate to the overall objective of the decision-making task — in this case maximising multilateral social welfare from the decision regarding allowing or preventing trade. Initially, criteria can be broad before being disaggregated into components or sub-criteria and even lower level sub-criteria. Around 7–12 of the lowest level of the criteria that can be quantitatively or qualitatively measured is regarded as ideal (Yoon and Hwang, 1995).

However, for illustrative purposes, let us assume that the Panel adopts three high-level criteria reflecting the three different forms the impacts on ecosystem services can take; regulating, provisioning and cultural. Further assume that each of these criteria are scored using a simple high–moderate–low categorisation according to the severity of ecosystem services impact that might result from trade. Example criteria descriptions are as follows:

- **Criterion 1: Regulating services**

Production of the good concerned might involve the destruction of ecosystems in the exporting country but might prevent the destruction of ecosystems in the importing country. The criteria description and scores are listed in Table 9.1.

- **Criterion 2: Provisioning services**

The impacts of trade on water, air and soil in both importing and exporting countries could affect the provision of services related to food and fibre, fresh

Table 9.1. Regulating services criterion.

Level	Description of Impact	Indicator
1	Trade is likely to a) cause low disruption to habitats, fire regimes, nutrient cycles or hydrological cycles; b) cause low disruption to the lifecycle of native species important to carbon sequestration, climate regulation, waste decomposition, water purification and pest and disease control.	Low
2	Trade is likely to a) cause moderate disruption to habitats, fire regimes, nutrient cycles or hydrological cycles; b) cause moderate disruption to the lifecycle of native species important to carbon sequestration, climate regulation, waste decomposition, water purification and pest and disease control.	Moderate
3	Trade is likely to a) cause high disruption to habitats, fire regimes, nutrient cycles or hydrological cycles; b) cause high disruption to the lifecycle of native species important to carbon sequestration, climate regulation, waste decomposition, water purification and pest and disease control.	High

Table 9.2. Provisioning services criterion.

Level	Description of Impact	Indicator
1	Trade has a low likelihood of affecting food, fresh water and genetic resources.	Low
2	Trade has a moderate likelihood of affecting food, fresh water and genetic resources.	Moderate
3	Trade has a high likelihood of affecting food, fresh water and genetic resources.	High

water and genetic diversity. The criterion description and scores are listed in Table 9.2.

- **Criterion 3: Cultural services**

Community stability and wellbeing could be affected by trade causing socio-economic cost or loss of biodiversity. For example, if the landed price of an imported good were sufficiently low to cause large segments of local

Table 9.3. Cultural services criterion.

Level	Impact Description	Indicator
1	Trade is likely to have a minor impact on public amenities, aesthetic values and biodiversity.	Low
2	Trade is likely to have a moderate impact on public amenities, aesthetic values and biodiversity.	Moderate
3	Trade is likely to have a high impact on public amenities, aesthetic values and biodiversity.	High

competitors to leave the market, local businesses, education and health services, sporting clubs and the like will experience a downturn. On the other hand, businesses that import the cheaper products could experience growth, expanding their businesses and labour and capital requirements. The criterion description and scores are listed in Table 9.3.

9.5.3. *Weighting criteria*

Decision-maker preferences in MCDA are reflected in weights placed on each of the criteria and sub-criteria. These weightings may place equal importance on all criteria, rank criteria from most to least important or assign a relative weighting to all criteria. The criteria weights may be qualitatively expressed, quantitatively expressed or a mixture of both. They can use direct estimation methods to determine the exact value of one criterion relative to another, or indirect estimation techniques that rely on weights used in past studies (Nijkamp *et al.*, 2013).

The weighting method chosen should be agreed to by the Panel during the dispute settlement process and need not necessarily be the same as a previous Panel so long as the process is open, transparent and agreeable to Panel members.

Let us assume in our example that the Panel feels each criterion should be assigned equal weighting. This might well be the most politically expedient weighting to adopt in practice to avoid conflict over the relative importance of cultural, provisioning and regulating ecosystem services.

9.5.4. *Assessment of the options*

The choice of which value the ecosystem penalty is assigned in a given case can be determined by evaluating each option in terms of each of the assessment criteria. The performance of each option can be displayed by

Table 9.4. Impact matrix.

Criteria	Ecosystem Services Penalty Option				
	1. High Cost	2. Moderate Cost	3. Low Cost	4. Moderate Benefit	5. High Benefit
Regulating ecosystem services	1	2	3	3	3
Provisioning ecosystem services	2	2	2	2	2
Cultural ecosystem services	1	2	3	2	1
Total	4	6	8	7	6

means of an *impact matrix* in which each of the components represents the evaluation of the options according to the individual criteria. An example is provided in Table 9.4.

The final ranking of each of the options is calculated by a mathematical operation using the impact matrix and criteria/sub-criteria weights. The form of this mathematical operation, referred to as the 'aggregation procedure', often describes the particular type of MCDA employed (Munda *et al.*, 1994). In our simplistic example, we simply use a weighted average that reveals option 3 to be the most agreeable option, so a 'low cost' ecosystem services penalty of 0% is the most preferred.

Note that all five options score the same against the provisioning ecosystem services criterion, meaning it is redundant in terms of distinguishing between options. If this were true for all three criteria, then the MCDA must be redesigned with more narrowly defined criteria so as to identify some differentiation between options. However, in Table 9.4, there is a disparity between option scores against the regulating ecosystem services and cultural ecosystem services criteria. It follows that any deliberative component of the MCDA should debate these two criteria and discard the provisioning ecosystem services criterion entirely.

9.5.5. *Sensitivity analysis*

Although not always undertaken, particularly in MCDAs that pursue a single optimal outcome, sensitivity analysis can be used to investigate the impact of uncertainty on the outcome of a particular analysis (Benke *et al.*, 2007). It can be conducted to reflect uncertainty in the estimation of values or weightings and if necessary can also consider the effects of different techniques used in the criteria weighting procedure. It thus adds to the overall understanding of complex problems and feeds directly into the iterative process that refines and improves the finally chosen option (Proctor and Drechsler, 2006).

Sensitivity of outcomes can be tested for different values of the most crucial and contentious criteria, weightings and impacts, helping to narrow the focus of the MCDA (Proctor and Drechsler, 2006). If, for example, there was a large disparity in preferences for one criterion, it could be enlightening to find out how the overall results change with different weights for this criterion. If the results are not greatly affected, then the criterion can take less importance in the overall process and the decision-makers can concentrate on other criteria. If the results are extremely sensitive to this criterion, then closer scrutiny should be given to it by confirming values and measurements.

9.5.6. *Deliberating, iterating and fine-tuning*

Practitioners of MCDA can achieve greater understanding of a decision-making problem by interacting with decision-makers to allow further iterations in the analysis if necessary. The initial list of options and criteria can be viewed as a starting point with the subsequent MCDA process contributing to the development of the most suitable options, criteria and weighting. In group decision-making situations like a Panel deciding on the value of an ecosystem services penalty, this step can be crucial if the ultimate aim of the analysis is to reach some compromise or agreement on the outcome.

It is important to recognize that making any risk assessment 'objective' when it involves non-market information adds complexity to the process and might not achieve freedom from bias. Even rigorous risk assessments inevitably reflect tacit yet dominant cultural values and identities, and are thus not completely objective, value-free processes (Slovic, 1999; Wynne, 1992). Introducing bureaucracies to enhance fairness and objectivity add to the intricacy of the process, which partly explains why the existing Dispute Settlement Process has proved to be complicated and lengthy (Davey, 2005; Mosoti, 2006). It is also expensive, meaning that high-income countries are the primary users of the system, although usage by countries with lower incomes has increased in recent years (Leitner and Lester, 2017).

The point is not whether MCDA would introduce subjective elements to dispute settlement, for these are already present. Rather, the issue is how they might be articulated and incorporated into a formal and structured decision analysis that is also practical (Keeney *et al.*, 1993). With this in mind, Deliberative Multi-Criteria Evaluation (DMCE), a particular form of MCDA useful in multi-faceted decision-making in group settings, could potentially form the base for future WTO Dispute Settlement. Introducing deliberation, reflection and adaption to MCDA, this form of decision facilitation could potentially transform Panel procedures into something more accessible and streamlined than the existing process.

DMCE combines the facilitation, interaction, and consensus-building features of jury processes with the structuring and integration features of traditional MCDA (Proctor and Drechsler, 2006). The jury involves a small group of participants charged with the responsibility of constituent representation and decision-making (Proctor and Drechsler, 2006). It is guided by an independent facilitator who ensures that participants have equal opportunity to express their views and the process is able to follow a suitable course to achieve a decision.

If utilised in WTO dispute settlement, DMCE using a traditional Dispute Settlement Panel as the jury could potentially lead to more streamlined process. Rather than Panel hearing testimony from opposing parties that may or may not be relevant to the decision of whether or not to allow trade, Members could instead call upon expert witnesses, technical analyses and anecdotal information to form individual opinions about issues critical to the decision of what ecosystem services penalty should apply to a particular case. Time might then be devoted to information clarification and discussion between Panel Members in which opinions are shared and modified in a process of group learning (Marinoni *et al.*, 2009).

9.6. Conclusion

In this chapter, we have discussed ways in which the WTO might explicitly consider ecosystem services impacts when ruling on trade disputes so as to internalise ecosystem services externalities, as well as the possible application of a consensus-building MCDA framework to help WTO Dispute Settlement Panels decide whether or not trade should go ahead with uncertain ecosystem services impact information. The semi-quantitative approach we have suggested by way of example is an extremely simplified version of a system that could be put in place to aid in the choice of an ecosystem services penalty which can then be incorporated into the decision to allow or prevent trade. The steps of the process we outlined are generic, applying to simple and complex MCDAs alike. Indeed, there need not necessarily be one prescribed MCDA framework that must be rigidly applied in all trade disputes but could instead allow a degree of flexibility so that different Panels can tailor the process according to the needs of specific cases.

We illustrated how the use of our simplified ecosystem services penalty system can be applied to the six cases in which combinations of multilateral gains from trade and multilateral ecosystem services impacts lead to different assessments of the desirability of trade, as identified in Chapter 7. Cases representing the greatest challenge from a dispute settlement perspective involve multilateral ecosystem services costs that exceed multilateral market benefits of trade and multilateral ecosystem services benefits from trade that exceed market losses. In both cases, a transparent, deliberative approach to decision-making such as provided by MCDA is called for in which Panel Members can utilise qualitative and quantitative information to help decide if trade should go ahead.

In Chapter 10, the final chapter, we summarise the contribution we think we have made in this book to analysing trade disputes, and on this basis suggest a way forward for improving WTO dispute resolution processes.

Chapter 10

Conclusion

10.1. Looking Back

Our motivation for writing this book was based on our perception of shortcomings in the GATT/WTO's existing procedures for evaluating and resolving international trade disputes which feature biosecurity and ecosystem services risks. Our view is that these procedures do not always deliver on their objective of enhancing social welfare but can be improved so that the consequent decisions in the context of environment-related trade disputes can consistently enhance social welfare.

Given this motivation, our broad objective in this book has been to demonstrate how the WTO's international trade dispute resolution procedures can be improved in terms of trade-related biosecurity and ecosystem services risks so as to enhance multilateral social, economic and environmental welfare.

This book has been structured into four parts. Part A, which comprised Chapters 2–4, contained a history and analysis of the GATT/WTO in the context of the resolution of environment-related international trade disputes. More specifically, Chapter 2 contained a brief history of the GATT/WTO, including Article XX and the Sanitary and Phytosanitary Standard (SPS) Agreement, and its trade dispute resolution procedures. Chapter 3 contained details and critical analysis of global trade dispute applications made through the SPS Agreement in the context of biosecurity risks. Finally, in Part A, Chapter 4 reviewed ecosystem services risk-related trade disputes through application of Article XX (the General Exceptions Article). As initially identified in Chapter 2 but confirmed by the detailed analysis of 18 trade dispute cases in Chapters 3 and 4, the WTO has continued to approach both trade–biosecurity and trade–ecosystem services disputes through interpretation of Article XX and the SPS Agreement with scant attention given to economic, environmental or social impacts of trade.

We contend in this book that in order to understand the holistic dispute under consideration, comprehensive data analysis of these impacts is required. There are many advantages to such an approach. First, it widens the scope of what is considered relevant to the decision and as such increases the chance that the Panel will determine an outcome in line with WTO objectives. Second, it enables quantification and qualification of the facts to be considered and therefore the ability to assess relatively between impacts. Previously, there have been practical barriers to implementing this approach, not least the ability to obtain meaningful data, to assess it based on rigorous valuation techniques but also in determining acceptable levels of uncertainty.

Following on from such limitations, Part B of this book, which comprised Chapters 5 and 6, demonstrated how a data-led approach to analysing biosecurity-related trade disputes is now both possible and beneficial to social welfare. Moreover, Part C of this book, which comprised Chapters 7 and 8, demonstrated how a data-led approach to analysing ecosystem services-related trade disputes is also now both possible and beneficial to social welfare.

More specifically, Part B built on the critique of existing WTO trade dispute procedures relating to biosecurity risks presented in Chapter 3 and focused on developing and applying a methodology for evaluating trade disputes which takes account of producer, consumer and biosecurity impacts in order to deliver a dispute resolution which enhances social welfare. In particular, Chapter 5 outlined the methodology developed in Cook and Fraser (2008), which both identifies and enables the evaluation of the key impacts of the disputed trade so that the best decision for social welfare can be made in relation to that trade. In addition, Chapter 6 then presented two case studies of the application of the methodology outlined in Chapter 5. These two case studies were drawn from those trade disputes identified in Chapter 3 and, while the quantification of these case studies was largely illustrative, Chapter 6 demonstrated in broad terms how the methodology of Chapter 5 can lead to different dispute resolutions than that produced by the existing rules-based approach of the WTO.

Specifically, in the case of NZ apples (DS367), the expected disease outbreak costs to domestic production associated with imports were larger than the estimated gains from allowing imports, implying that allowing trade results in a net loss of social welfare and thereby contradicting the WTO ruling. Whereas in the Canadian salmon case (DS18), the predicted gains from trade were larger than the expected disease costs, implying that allowing trade results in a net gain in social welfare and thereby endorsing the WTO ruling. On this basis we argued that analytical frameworks where the benefits of trade to consumers and bioeconomic modelling of invasive species spread and control are reconciled are needed to solve import trade disputes in the best interest of social welfare.

Part C built on the critique of existing WTO trade dispute procedures relating to ecosystem services risks presented in Chapter 4 and focused on developing and applying a methodology for evaluating trade disputes which takes account of producer, consumer and ecosystem services impacts in order to deliver a dispute resolution which enhances social welfare. In particular, Chapter 7 outlined the methodology developed in Cook and Fraser (2015), which identified and then considered evaluating the key impacts of the disputed trade with a view to making the best decision for global welfare in relation to that trade. In addition, Chapter 8 then presented two case studies of the application of the methodology outlined in Chapter 7. These two case studies were drawn from those trade disputes identified in Chapter 4 and, once again, while the quantification of these case studies was largely illustrative, Chapter 8 demonstrated in broad terms how the methodology of Chapter 7 can lead to different dispute resolutions than that produced by the existing rules-based approach of the WTO.

While for both case studies, the actual WTO ruling was to support trade, the analysis in this chapter suggested that in the Brazilian/EU retreaded tyre case study (DS332), the estimated ecosystem services damages prevented from banning imported EU retreaded tyres were larger than the traditional gains from trade forfeited, implying that allowing trade results in a net loss of social welfare and thereby contradicting the WTO ruling. Whereas in the case of the US/Mexican tuna dispute the estimated prevented ecosystem services damages from banning imported Mexican tuna (ILM1594) were smaller than the traditional gains from trade lost as a result of the ban, implying that allowing trade results in a net gain in social welfare and thereby supporting the WTO ruling. On this basis we argued that analytical frameworks which take account of producer, consumer and ecosystem services impacts are needed to solve import trade disputes in the best interest of social welfare.

Note that the central difference between Parts B and C was that in the case of Part B the impacts were largely economic, and therefore lend themselves to standard economic evaluation whereas in Part C the impacts were largely social and environmental, and therefore typically non-monetised, which created a different evaluation problem to be solved. This provided the motivation for Chapter 9 in which we discussed ways in which the WTO might explicitly consider ecosystem services impacts when ruling on trade disputes so as to internalise ecosystem services externalities. In addition, we considered the possible application of a consensus-building MCDA framework to help WTO Dispute Settlement Panels decide whether or not trade should go ahead with uncertain ecosystem services impact information. In so doing, we illustrated how the use of an ecosystem services penalty system can be applied to the six cases identified in Chapter 7 in which combinations of multilateral gains from trade and multilateral ecosystem services impacts led to different assessments of the desirability of trade. In all cases, a transparent, deliberative approach to decision-making such as provided by the use of MCDA is called for in which Panel Members can utilise qualitative and quantitative information to help decide if trade should go ahead.

10.2. Looking Forward

We have found that previous literature in the area of WTO trade disputes has not taken a particular interest in environment-related trade disputes, and what literature we have found typically adopts a 'descriptive' approach to presenting relevant content, rather than an analysis/evaluation approach which we advocate here as being required to make social welfare-enhancing trade dispute decisions. On this basis we present our case for modifying the WTO's existing approach to international trade dispute resolution in the context of biosecurity and ecosystem services risks so that their decisions can consistently enhance social welfare.

So, what are the likely barriers to the future adoption of this approach by the WTO?

We think there are three main impediments:

(1) **The motivation of the WTO to change its current dispute resolution process.**
As identified in Chapters 2–4, the current WTO dispute resolution process can best be described as 'rule-based'. This description is consistent with the clear dominance of legal technicalities within the dispute resolution process and reflects the control of the current process

by legal experts. In addition, we noted in Chapters 3 and 4 that, even where scientific data relating to trade impacts was provided to the Dispute Resolution Panel, this data was typically ignored by the Panel in making its judgement. Moreover, we noted that in the few cases where economic analysis of trade impacts was provided to the Panel, this information was also ignored by the Panel in making its judgement.

It follows that the Dispute Resolution Panel is likely to be resistant to a new requirement being imposed on it to take account of an evaluation of the economic, environmental and social impacts of trade as part of the dispute resolution process. Nevertheless, given the size of the enduring benefits to multilateral social welfare from making the correct decision regarding allowing or preventing trade estimated for the case studies in Chapters 6 and 8, it seems reasonable to hope that the Panel would appreciate the merits of changing its current process.

(2) **A more costly dispute resolution process.**
Even though economic research is typically viewed as 'low cost' relative to scientific research, there seems little doubt that for biosecurity-related cases adding the sort of evaluation feature, we have proposed to the existing WTO dispute resolution process could be seen to increase its cost. In the case of ecosystem services-related cases, this increase in cost could be larger if the MCDA process to evaluate ecosystem services impacts is also included. However, the approach suggested in this book, if used *as an alternative* to the current process whereby plaintiffs and defendants can submit any information they feel may be appropriate, may actually be less costly. In particular, sifting through such a flood of data to determine whether it has any *prima facie* relevance to the case is in itself a costly process. Asking for data in a targeted way, then analysing it, should provide a more cost-effective solution.

Significantly, any added costs are unlikely to exceed several million dollars, and therefore are likely to be only a small percentage of the in-perpetuity added benefits to multilateral social welfare from making the correct decision regarding preventing or allowing trade — which as suggested by the case studies in Chapters 6 and 8 are estimated to vary from \$30 million to almost \$1 billion over a 20-year period.

(3) **The role of the WTO in international trade agreements in the future.**
Since its creation in 1995, the WTO has struggled to move forward its agenda of multilateral trade liberalisation. While a part of this struggle can be attributed to anti-free trade political sentiments, perhaps a more

important impediment has been the massive expansion in the number and value of bilateral trade agreements established in the last two decades.[1] This expansion is evidence that countries have been finding the process of liberalising trade on a bilateral basis, with agreements that are outside the WTO framework, more attractive than working towards this goal on a multilateral basis within the WTO framework.

A related concern in this context is that of corporate action. The corporate world appears ready to take advantage of this new-found interest in agreements outside of the WTO, potentially worsening the neglect of ecosystem services impacts in trade decisions.[2]

If such initiatives take off elsewhere it follows that the future role of the WTO in international trade agreements may be increasingly sidelined. This could make the WTO itself less inclined to be innovative and therefore increasingly irrelevant. Our hope, however, is that with an improvement in its dispute resolution procedures as proposed in this book, the appeal will be enhanced for countries once again to work towards WTO-based multilateral trade agreements, not least as this approach provides a process whereby trade decisions are made by optimising social welfare.

[1]For example, the top 12 bilateral agreements (in terms of value of trade) established between the US and another country have all been since the year 2000. See https://www.thebalance.com/what-are-bilateral-trade-agreements-pros-cons-and-list-3305911 (viewed 05/03/2019).

[2]In a recent example, the Comprehensive and Progressive Agreement for Trans-Pacific Partnership (CPTPP) signed by 11 countries (Australia, Brunei, Canada, Chile, Japan, Malaysia, Mexico, New Zealand, Peru, Singapore and Vietnam) which came into effect on 30 December 2018 inhibits the ability of democratic political institutions to invoke policy and regulations that affect transnational capital. By circumventing over 18,000 government regulations, including environmental standards, the CPTPP aims to create a unified trade and investment environment that delivers corporate income gains (Chodor, 2019). Unlike the WTO framework, however, the signatories agreed that no draft texts or working documents were to be released during negotiations, and no documents other than the final text released until 4 years after the agreement took effect, preventing critical commentary and analysis (Kelsey, 2013).

References

ABARE–BRS. 2010. Australian fisheries statistics 2009. Canberra, Australian Bureau of Agricultural and Resource Economics — Bureau of Rural Sciences.

Abelson, P. 2007. Establishing a monetary value for lives saved: Issues and controversies. Working Papers in Cost benefit Analysis WP 2008-2. Department of Finance and Deregulation.

ABS. 2004. AgStats Integrated Regional Database, Catalogue No. 1353.0. Canberra, Australia: Australian Bureau of Statistics.

ADVS. 1999. Final Report. Consultancy on routes for exposure of aquatic animals to aquatic animal products intended for human consumption. Prepared for the Australian Quarantine and Inspection Service. Allens Rivulet, Tasmania: Aquaculture Development and Veterinary Services Pty. Ltd.

AHA. 2002. *Government and livestock industry cost sharing deed in respect of emergency animal disease response.* Canberra, Australia: Animal Health Australia.

Andersen, T. B., K. H. Roll, and S. Tveterås. 2008. The price responsiveness of salmon supply in the short and long run. *Marine Resource Economics* 23: 425–437.

Andow, D. A., P. Kareiva, S. A. Levin, and A. Okubo. 1990. Spread of invading organisms. *Landscape Ecology* 4: 177–188.

Australian Government Department of Agriculture Fisheries and Forestry. 2011. *Import risk analysis handbook.* Canberra, Australia: Australian Government Department of Agriculture, Fisheries and Forestry.

Australian Quarantine and Inspection Service. 1995. Import risk analysis — disease risks associated with the importation of uncooked, wild, ocean caught Pacific Salmon product from the USA and Canada. Canberra, Australia: Australian Quarantine and Inspection Service.

Australian Quarantine and Inspection Service. 1999. Import risk assessment on non-viable salmonids and non-salmonid marine finfish. Canberra, Australia: Australian Quarantine and Inspection Service.

Babula, R. A., and R. L. Corey. 2004. U.S. canned tuna supply and demand. *Journal of International Food & Agribusiness Marketing* 16: 145–164.

Barbier, E. B., I. Strand, and S. Sathirathai. 2002. Do open access conditions affect the valuation of an externality? Estimating the welfare effects of mangrove-fishery linkages in Thailand. *Environmental and Resource Economics* 21: 343–365.

Baughen, S. 2007. *International Trade and Protection of the Environment.* Abingdon, Oxon, UK: Routledge-Cavendish.

BBC News. 2015. Brazil faces surge in number of dengue fever cases [Online]. BBC. Available at: https://www.bbc.com/news/world-latin-america-32589268 [Accessed 26th October 2018].

Benke, K. K., A. J. Hamilton, and K. E. Lowell. 2007. Uncertainty analysis and risk assessment in the management of environmental resources. *Australasian Journal of Environmental Management* 14: 243–249.

Bhati, U. N., and C. Rees. 1996. *Fireblight: A Cost Analysis of Importing Apples from New Zealand.* Canberra, Australia: Australian Bureau of Agricultural and Resource Economics.

Biosecurity Australia. 2001. *Guidelines for Import Risk Analysis.* Canberra, Australia: Agriculture, Fisheries and Forestry Australia/Biosecurity Australia.

Biosecurity Australia. 2004. Importation of apples from New Zealand — Revised draft IRA Report. Department of Agriculture, Fisheries and Forestry.

Blandford, D. 2011. International trade policies and ecosystem services. In Koellner, T., ed., *Ecosystem Services and Global Trade of Natural Resources.* New York, NY: Routledge.

Britton, N. F. 1986. *Reaction-Diffusion Equations and Their Applications to Biology.* London: Academic Press.

Brown, J. 2005. An account of the dolphin-safe tuna issue in the UK. *Marine Policy* 29: 39–46.

Brown, C. P., and J. P. Trachtman. 2009. Brazil — Measures affecting imports of retreaded tyres: A balancing act. In Horn, H., and P. C. Mavroidis, eds., *The WTO Case Law of 2006–7.* Cambridge, UK: Cambridge University Press.

Bryan, R. 2018. Trump's trade war with China is the biggest threat to the US economy in 2019, and it's making economists the most worried they have been in years. Business Insider Australia [Online]. Available at: https://www.businessinsider.com.au/trump-china-trade-war-tariffs-biggest-2019-us-economic-threat-2018-12?r=US&IR=T [Accessed 23rd January 2019].

Buckner, S. 1995. Reducing fireblight. *The Grower.* October: 8.

Burgman, M., F. Fidler, M. McBride, T. Walshe, and B. Wintle. 2006. Eliciting expert judgements: Literature review. Melbourne, Australia: Australian Centre of Excellence for Risk Analysis.

Burke, M., K. Oleson, E. McCullough, and J. Gaskell. 2009. A global model tracking water, nitrogen, and land inputs and virtual transfers from industrialized meat production and trade. *Environmental Modeling & Assessment* 14: 179–193.

Burton, S. J. 1995. An Introduction to Law and Legal Reasoning, second edition. Boston, MA: Little, Brown and Company.

CABI/EPPO. 2003. *Crop Protection Compendium — Global Module.* Cayman Islands: CAB International.

Cameron, J., and H. Ward. 1994. The mulitlateral trade organisation: A revised perspective. In Cameron, J., P. Demeret, and D. Geradin, eds., *Trade and the Environment: A Search for Balance*. London: Cameron May.

Carey, J. M., and M. A. Burgman. 2008. Linguistic uncertainty in qualitative risk analysis and how to minimize it. *Annals of the New York Academy of Sciences* 1128: 13–17.

Charnovitz, S. 1993. Environmental harmonisation and trade policy. In Zaelke, D., P. Orbuch, and R. F. Housman, eds., *Trade and the Environment: Law, Economics and Policy*. Washington DC, USA: Center for International Environmental Law, Island Press.

Cheng, H.-T., and O. Capps, Jr. 1988. Demand analysis of fresh and frozen finfish and shellfish in the United States. *American Journal of Agricultural Economics* 70: 533–542.

Chodor, T. 2019. The rise and fall and rise of the trans-pacific partnership: 21st century trade politics through a new constitutionalist lens. *Review of International Political Economy* 26: 232–255.

Commonwealth of Australia. 2006. *Handbook of Cost-Benefit Analysis*. Canberra, Australia: Department of Finance and Administration.

Cook, D. C. 2003. Prioritising exotic pest threats to Western Australian plant industries. Discussion Paper. Bunbury: Government of Western Australia — Department of Agriculture.

Cook, D. C. 2008. Benefit cost analysis of an import access request. *Food Policy* 33: 277–285.

Cook, D. C., and R. W. Fraser. 2008. Trade and invasive species risk mitigation: Reconciling WTO compliance with maximising the gains from trade. *Food Policy* 33: 176–184.

Cook, D. C, and R. W. Fraser. 2015. Ecosystem services as a catalyst for multilateral net welfare maximising trade decisions. Paper presented to the Annual Conference of the Agricultural Economics Society, April, University of Warwick.

Costanza, R., R. D'Arge, R. De Groot, S. Farber, M. Grasso, B. Hannon, K. Limburg, S. Naeem, R. V. O'Neill, J. Paruelo, R. G. Raskin, P. Sutton, and M. Van der Belt. 1997. The value of the world's ecosystem services and natural capital. *Nature* 387: 253–260.

Costello, C., and C. McAusland. 2003. Protectionism, trade, and measures of damage from exotic species introductions. *American Journal of Agricultural Economics* 85: 964–975.

Cramer, K. L., W. L. Perryman, and T. Gerrodette. 2008. Declines in reproductive output in two dolphin populations depleted by the yellowfin tuna purse-seine fishery. *Marine Ecology Progress Series* 369: 273–285.

Davey, W. J. 2005. The WTO dispute settlement system: The first ten years. *Journal of International Economic Law* 8, 17–50.

Department of Agriculture and Water Resources. 2016. *Biosecurity Import Risk Analysis Guidelines 2016: Managing Biosecurity Risks for Imports into Australia*. Canberra, Australia: Department of Agriculture and Water Resources.

DPIE. 1988. Australian quarantine requirements for the future: Report of the Quarantine Review Committee. Canberra, Australia: Department of Primary Industries and Energy.

Dwyer, G. 1992. On the spatial spread of insect pathogens — theory and experiment. *Ecology* 73: 479–494.

Eiswerth, M. E., and W. S. Johnson. 2002. Managing nonindigenous invasive species: Insights from dynamic analysis. *Environmental & Resource Economics* 23: 319–342.

Ekins, P., C. Folke, and R. Costanza. 1994. Trade, environment and development: The issues in perspective. *Ecological Economics* 9: 1–12.

Engel, S., S. Pagiola, and S. Wunder. 2008. Designing payments for environmental services in theory and practice: An overview of the issues. *Ecological Economics* 65: 663–674.

Esty, D. 1994. *Greening the GATT: Trade, Environment and the Future.* Washington, USA: Institute for International Economics, USA.

FAO. 2009. *FAOSTAT Database.* Rome, Italy: Food and Agriculture Organization of the United Nations.

Finnoff, D., J. F. Shogren, B. Leung, and D. Lodge. 2007. Take a risk: Preferring prevention over control of biological invaders. *Ecological Economics* 62: 216–222.

Finnoff, D., and J. Tschirhart. 2005. Identifying, preventing and controlling invasive plant species using their physiological traits. *Ecological Economics* 52: 397–416.

Fisher, R. A. 1937. The wave of advance of advantageous genes. *Annual Eugenics* 7: 353–369.

Francis, R. C., F. T. Awbrey, C. L. Goudey, M. A, Hall, D. M. King, H. Medina, K. S. Norris, M. K. Orbach, R. Payne, and E. Pikitch. 1992. *Dolphins and the Tuna Industry.* Washington, DC, USA: National Research Council, National Academy Press.

Freeman III, A. M. 2002. How much is Nature really worth? An economic perspective. Valuing nature: the Shipman Workshop papers. Brunswick, Maine: Bowdoin College, Environmental Studies Program.

Galarraga, I., and A. Markandya. 2003. The analysis of welfare effects of an environmental product tax: An application to the taxation of car tyres in Hungary. *Fiscal Studies* 24, 97–114.

Galloway, J. N., M. Burke, G. E. Bradford, R. Naylor, W. Falcon, A. K. Chapagain, J. C. Gaskell, E. McCullough, H. A. Mooney, K. L. L. Oleson, H. Steinfeld, T. Wassenaar, and V. Smil. 2007. International trade in meat: The tip of the pork chop. *Ambio* 36: 622–629.

GATT. 1991. United States — Restrictions on imports of tuna, Report of the Panel, DS21/R — 39S/155 (September 3). Geneva: General Agreement on Tariffs and Trade.

GATT. 1994. Agreement on the application of Sanitary and Phytosanitary Measures. The results of the Uruguay Round of Multilateral Trade Negotiations: The legal texts. Geneva: General Agreement on Tariffs and Trade Secretariat.

Goldstein, J. L., D. Rivers, and M. Tomz. 2007. Institutions in international relations: Understanding the effects of the GATT and the WTO on World Trade. *International Organization* 61(1): 37–67.

Gray, K. R. 2008. Brazil: Measures affecting imports of retreaded tyres. *The American Journal of International Law* 102: 610-616.

Gutierrez, A. P., and U. Regev, U. 2005. The bioeconomics of tritrophic systems: Applications to invasive species. *Ecological Economics* 52: 383–396.

Haddock, J. 2000. Rule-based decision making and its applicability to environmental-related trade disputes in the World Trade Organisation (WTO). *Journal of Environmental Assessment and Policy Management* 2 (1 March): 81–98.

HAL. 2004. *The Australian Horticulture Statistics Handbook.* Sydney, Australia: Horticulture Australia Limited.

Hall, M., and M. Roman. 2013. Bycatch and non-tuna catch in the tropical tuna purse seine fisheries of the world. FAO Fisheries and Aquaculture Technical Paper No. 568. Rome, Italy: Food and Agriculture Organization of the United Nations.

Haraden, J., S. F. Herrick, D. Squires, and C. A. Tisdell. 2004. Economic benefits of dolphins in the United States Eastern Tropical Pacific Purse-Seine Tuna Industry. *Environmental and Resource Economics* 28: 451–468.

Hengeweld, R. 1989. *The Dynamics of Biological Invasions.* London, UK: Chapman and Hall.

Hinchy, M. D., and J. Low. 1990. Cost-benefit analysis of quarantine regulations to prevent the introduction of fire blight into Australia: Report to the Australian Quarantine and Inspection Service. Canberra, Australia: Australian Bureau of Agricultural and Resource Economics.

Hoekstra, A. Y., and P. Q. Hung. 2005. Globalisation of water resources: International virtual water flows in relation to crop trade. *Global Environmental Change* 15: 45–56.

Holmes, E. E. 1993. Are diffusion-models too simple – a comparison with telegraph models of invasion. *American Naturalist* 142: 779–795.

Horan, R. D., C. Perrings, F. Lupi, and E. H. Bulte. 2002a. Biological pollution prevention strategies under ignorance: The case of invasive species. *American Journal of Agricultural Economics* 84, 1303–1310.

Horan, R. D., C. Perrings, F. Lupi, and E. H. Bulte. 2002b. The economics of invasive species management: Biological pollution prevention strategies under ignorance: The case of invasive species. *American Journal of Agricultural Economics* 84: 1303–1310.

Hudec, R. E. 1990. *The GATT Legal System and World Trade Diplomacy.* Salem, NH, USA: Butterworth Legal Publishers.

ICTSD. 2010. Litigating environmental protection and public health at the WTO: The Brazil-retreaded tyres case. ICTSD information note no. 1. San Francisco: International Centre for Trade and Sustainable Development.

Industry Commission. 1996. *Australian Atlantic Salmon: Effects of Import Competition.* Canberra, Australia: Industry Commission.

Inter-American Tropical Tuna Commission. 1989. Incidental mortality of dolphins in the eastern tropical pacific fishery, 1979–1988. A decade of the Inter-American Tropical Tuna Commission's Scientific Technicians Program. Working Document No. 2 Tuna-Dolphin Workshop, March 14–16. San Jose, Costa Rica: Scripps Institution of Oceanography.

Jankowski, P., and T. Nyerges. 2001. GIS-supported collaborative decision-making: Results of an experiment. *Annals of the Association of American Geographers* 91, 48–70.

Joyner, C. C., and Z. Tyler. 2000. Marine conservation versus international free trade: Reconciling dolphins with tuna and sea turtles with shrimp. *Ocean Development & International Law* 31, 127–150.

Kareiva, P., and M. Marvier. 2003. Conserving biodiversity coldspots. *American Scientist* 91, 344–351.

Kaul, S., K. J. Boyle, N. V. Kuminoff, C. F. Parmeter, and J. C. Pope. 2013. What can we learn from benefit transfer errors? Evidence from 20 years of research on convergent validity. *Journal of Environmental Economics and Management* 66, 90–104.

Kazana, V., L. Sisak, H. Jobstl, and M. Merlo. 2000. A critical overview of multiple criteria decision making methods for natural resource planning and management with reference to practice. In H. Jöbstl, M. Merlo, and L. Sisak, (eds.) *Proceedings of the International Union of Forest Research Organizations Symposium 'From Theory to Practice: Gaps and Solutions in Managerial Economics and Accounting in Forestry', Prague, 13–15 May 1999.* Prague: Czech University of Agriculture.

Keeney, R. L., H. Raiffa, and R. F. Meyer. 1993. *Decisions with Multiple Objectives: Preferences and Value Tradeoffs.* Cambridge, UK: Cambridge University Press.

Kelsey, J. 2013. The Trans-Pacific Partnership Agreement: A gold-plated gift to the global tobacco industry? *American Journal of Law & Medicine* 39: 237–264.

Kim, C. S., R. N. Lubowski, J. Lewandrowski, and M. E. Eiswerth. 2006. Prevention or control: Optimal government policies for invasive species management. *Agricultural and Resource Economics Review* 35: 29–40.

Kirchhoff, S., B. G. Colby, and J. T. LaFrance. 1997. Evaluating the performance of benefit transfer: An empirical inquiry. *Journal of Environmental Economics and Management* 33: 75–93.

Knowler, D. 2005. Reassessing the costs of biological invasion: *Mnemiopsis leidyi* in the Black sea. *Ecological Economics* 52: 187–199.

Koellner, T., ed. 2011. *Ecosystem Services and Global Trade of Natural Resources.* London, UK and New York, NY: Routledge.

Koellner, T., S. Pfister, and A. Koehler. 2011. Life cycle assessment and ecosystem services. In Koellner, T., ed., *Ecosystem Services and Global Trade of Natural Resources.* London, UK and New York, NY: Routledge.

Langlois, R. N., N. J. and Foss, 1999. Capabilities and governance: The rebirth of production in the Theory of Economic Organization. *Kyklos* 52: 201–218.

Lavranos, N., and N. Vielliard. 2008. Brazilian tyres case: The competing trade and non-trade interests and competing jurisdictions between MERCOSUR and WTO. *European Energy and Environmental Law Review* 17: 306–318.

League of Nations Secretariat. 1945. Commercial policy in the post-war world, Annex III (Official number C.31.1945.II.A), pp. 118–124, Geneva, Switzerland: League of Nations.

Leitner, K., and S. Lester. 2017. WTO Dispute Settlement 1995–2016: A statistical analysis. *Journal of International Economic Law* 20: 171–182.

Levine, J. M., and C. D'Antonio. 2003. Forecasting biological invasions with increasing international trade. *Conservation Biology* 17, 322–326.

Lewis, M. A. 1997. Variability, patchiness, and jump dispersal in the spread of an invading population. In Tilman, D., and P. Kareiva, eds., *Spatial Ecology: The Role of Space in Population Dynamics and Interspecific Interactions.* New Jersey, NJ: Princeton University Press.

Liu, S., R. Costanza, A. Troy, J. D'Aagostino, and W. Mates. 2010a. Valuing New Jersey's ecosystem services and natural capital: A spatially explicit benefit transfer approach. *Environmental Management* 45, 1271–1285.

Liu, S., W. Proctor, and D. Cook. 2010b. Using an integrated fuzzy set and deliberative multi-criteria evaluation approach to facilitate decision-making in invasive species management. *Ecological Economics* 69: 2374–2382.

Mack, R., and W. Lonsdale. 2001. Humans as global plant dispersers: Getting more than we bargained for. *Bioscience* 51: 95–102.

Mack, R. N. 1996. Predicting the identity and fate of plant invaders: Emergent and emerging approaches. *Biological Conservation* 78: 107–121.

Marinoni, O., A. Higgins, S. Hajkowicz, and K. Collins. 2009. The multiple criteria analysis tool (MCAT): A new software tool to support environmental investment decision making. *Environmental Modelling & Software* 24: 153–164.

Markiw, M. E., and K. Wolf. 1983. *Myxosoma cerebralis* (Myxozoa: Myxosporea) etiologic agent of salmonid whirling disease requires tubificid worm (Annelida: Oligochaeta) in its life cycle. *The Journal of Protozoology* 30, 561–564.

McCann, K., A. Hastings, S. Harrison, and W. Wilson. 2000. Population outbreaks in a discrete world. *Theoretical Population Biology* 57: 97–108.

McDorman, T. L. 1992. The 1991 US-Mexico GATT Panel Report on tuna and dolphin: Implications for trade and environment conflicts. *North Carolina Journal of International Law and Commercial Regulation* 17: 461.

McElwee, H. 2000. Economic impact assessment of fireblight (*Erwinia Amylovora*) in Western Australia. Unpublished Draft Report. Bunbury, Australia: Government of Western Australia – Department of Agriculture.

McKelvie, L. 1991. The economic impact of whirling disease on the Australian salmonid industry. Canberra, Australia: Australian Bureau of Agricultural and Resource Economics.

McKelvie, L., C. Reid, and M. Haque. 1994. Economic impact of salmonid diseases: Furunculosis and infectious haematopoeitic necrosis (IHN): Report to the Australian Quarantine and Inspection Service. Canberra, Australia: Australian Bureau of Agricultural and Resource Economics.

Millennium Ecosystem Assessment. 2005. *Ecosystems and Human Well-Being: Synthesis.* Washington, D.C.: Island Press.

Miller, T. R. 2000. Variations between countries in values of statistical life. *Journal of Transport Economics and Policy* 34: 169–188.

Moody, M. E., and R. N. Mack. 1988. Controlling the spread of plant invasions: The importance of nascent foci. *Journal of Applied Ecology* 25: 1009–1021.

Mosoti, V. 2006. Africa in the first decade of WTO dispute settlement. *Journal of International Economic Law* 9: 427–453.

Munda, G., P. Nijkamp, and P. Rietveld. 1994. Qualitative multicriteria evaluation for environmental management. *Ecological Economics* 10: 97–112.

Murphy, D. D. 2006. The tuna-dolphin wars. *Journal of World Trade* 40: 597–617.

Musgrave, R. K., and G. Stephens. 1993. The GATT-tuna dolphin dispute: An update. *Natural Resources Journal* 33: 957–975.

Nelson, R., M. Howden, and M. Stafford-Smith. 2008. Using adaptive governance to rethink the way science supports Australian drought policy. *Environmental Science & Policy* 11: 588–601.

Nijkamp, P., P. Rietveld, and H. Voogd. 2013. *Multicriteria Evaluation in Physical Planning.* Amsterdam, North-Holland.

O'Rourke, D. 2007. *World Apple Review.* Pullman, WA: Belrose Inc.

OECD 2018. *Agricultural Policy Monitoring and Evaluation 2018.* Paris, France: OECD Publishing.

Office of Best Practice Regulation. 2014. *Best practice regulation guidance note: Value of statistical life.* Canberra, Australia: Department of Prime Minister and Cabinet.

Okubo, A. and S. A. Levin. 2002. *Diffusion and Ecological Problems: Modern Perspectives.* New York, NY: Springer.

Olson, L., and S. Roy. 2005. On prevention and control of an uncertain biological invasion. *Review of Agricultural Economics* 27: 491–497.

Olson, L. J., and S. Roy. 2002. The economics of controlling a stochastic biological invasion. *American Journal of Agricultural Economics* 84: 1311–1316.

Paini, D. R., S. P. Worner, D. C. Cook, P. J. D. Barro, and M. B. Thomas. 2010. Using a self-organizing map to predict invasive species: sensitivity to data errors and a comparison with expert opinion. *Journal of Applied Ecology* 47: 290–298.

Parker, R. W. 1999. The use and abuse of trade leverage to protect the global commons: What we can learn from the tuna-dolphin conflict. *Georgetown International Environmental Law Review* 12: 1.

Perrings, C., M. H. Williamson, and S. Dalmazzone, eds. 2000. *The Economics of Biological Invasions.* Cheltenham, UK: Edward Elgar Publishing.

Peters, G. P., and E. G. Hertwich. 2008. CO_2 embodied in international trade with implications for global climate policy. *Environmental Science & Technology* 42: 1401–1407.

PHA. 2001. Funding and compensation for emergency eradication of exotic plant pests and diseases: A Discussion Paper. Canberra, Australia: Plant Health Australia.

PHA. 2005. *Government and Plant Industry Cost Sharing Deed in Respect of Emergency Plant Pest Responses.* Canberra, Australia: Plant Health Australia.

Plummer, M. L. 2009. Assessing benefit transfer for the valuation of ecosystem services. *Frontiers in Ecology and the Environment* 7: 38–45.

Proctor, W., and M. Drechsler. 2006. Deliberative multicriteria evaluation. *Environment and Planning C: Government and Policy* 24: 169–190.

Ready, R., and S. Navrud. 2006. International benefit transfer: Methods and validity tests. *Ecological Economics* 60: 429–434.

Ready, R., S. Navrud, B. Day, R. Dubourg, F. Machado, S. Mourato, F. Spanninks, and M. X. V. Rodriquez. 2004. Benefit transfer in Europe: How reliable are transfers between countries? *Environmental and Resource Economics* 29: 67–82.

Rietveld, P. 1980. *Multiple Objective Decision Methods and Regional Planning.* Amsterdam, North-Holland.

Roberts, W. 1991. Consequences of establishment of fireblight in Australia, Information Paper No. IP/1/91. Canberra, Australia: Bureau of Rural Resources.

Rosenberger, R. S., and T. D. Stanley. 2006. Measurement, generalization, and publication: Sources of error in benefit transfers and their management. *Ecological Economics* 60: 372–378.

Saito, K. 2001. Yardsticks for "Trade and Environment": Economic analysis of the WTO Panel and Appellate Body reports regarding environment-orientated trade measures, Jean Monnet Working Paper 14/01. The Jean Monnet Program.

Sakagawa, G. T. 1991. Are U.S. regulations on tuna–dolphin fishing driving U.S. seiners to foreign-flag registry. *North American Journal of Fisheries Management* 11: 241–252.

Schaefer, M. B. 1957. Some considerations of population dynamics and economics in relation to the management of the commercial marine fisheries. *Journal of the Fisheries Research Board of Canada* 14, 669–681.

Scott, M. D., S. J. Chivers, R. J. Olson, P. C. Fiedler, and K. Holland. 2012. Pelagic predator associations: Tuna and dolphins in the eastern tropical Pacific Ocean. *Marine Ecology Progress Series* 458: 283–302.

Seale, J. L., A. Regmi, and J. Bernstein. 2003. International evidence on food consumption patterns. Technical Bulletin Number 1904. Washington, DC: Economic Research Service, United States Department of Agriculture.

Shigesada, N., and K. Kawasaki. 1997. *Biological Invasions: Theory and Practice.* Oxford, UK: Oxford University Press.

Shutt, H. 1985. *The Myth of Free Trade.* London, UK: Basil Blackwell Ltd and The Economist Publications Ltd.

Skellam, J. G. 1951. Random dispersal in theoretical populations. *Biometrika* 38: 196–218.

Slovic, P. 1999. Trust, emotion, sex, politics, and science: Surveying the risk-assessment battlefield (Reprinted from Environment, ethics, and behavior, pg 277–313, 1997). *Risk Analysis* 19: 689–701.

Snape, R. H., and D. Orden. 2012. Integrating import risk and trade benefit analysis. In Anderson, K., C. McRae, and D. Wilson, eds., *The Economics of Quarantine and the SPS Agreemen.*University of Adelaide Press, Adelaide, South Australia, pp. 174–182. [First published in 2001 by Centre for International Economic Studies and the Department of Agriculture, Fisheries and Forestry – Australia/Biosecurity Australia].

Stanton, G. H. 2012. The WTO dispute settlement framework and operation. In Anderson, K., C. McRae, and D. Wilson, eds. *The Economics of Quarantine and the SPS Agreement.* University of Adelaide Press, Adelaide, South Australia, pp. 53–74. [First published in 2001 by Centre for International Economic Studies and the Department of Agriculture, Fisheries and Forestry — Australia/Biosecurity Australia].

Thaggert, H. L. 1994. A closer look at the tuna-dolphin case: "Like Products" and "Extrajurisdictionality" in the trade and environment context. In Cameron, J., P. Damaret, and D. Geradin, eds., *Trade and the Environment: The Search for Balance.* London, UK: Cameron May.

Thomson, A. P. 2001. Australia-salmon and compliance issues surrounding the SPS agreement: sovereign acceptance and measure adaptation. *Law and Policy in International Business* 33: 717.

Trebilcock, M.J., and R. Howse. 1995. *The Regulation of International Trade.* London, UK: Routledge.

Valdes, A., and J. Zietz. 1980. Agricultural protection in OECD countries: Its cost to less developed countries. IFPRI Research Report 21. Washington, D.C.: International Food Policy Research Institute.

Van Beukering, P. J. H., and M. A. Janssen. 2001. Trade and recycling of used tyres in Western and Eastern Europe. *Resources, Conservation and Recycling* 33: 235–265.

Van Damme, I. 2008. Appellate Body Report, Brazil — Measures affecting imports retreaded tyres, adopted on 17 December 2007. *International and Comparative Law Quarterly* 57: 710–723.

Vandevelde, K. J. 1996. *Thinking Like a Lawyer: An Introduction to Legal Reasoning.* New York, NY: Westview Press.

Viljoen, J., M. McGillivray, T. Orton, and G. Oliver. 1997. *The Potential Impact of Fire Blight on the Australian Apple and Pear Industry: A Socio-Economic Study.* Camberwell, UK: Corporate Strategy Consulting.

Viscusi, W. K. 1993. The value of risks to life and health. *Journal of Economic Literature* 31: 1912–1946.

Vose, D. 2000. *Risk Analysis: A Quantitative Guide.* Chichester, UK: John Wiley & Sons.

Waage, J. K., R. W. Fraser, J. D. Mumford, D. C. Cook, and A. Wilby. 2005. *A New Agenda for Biosecurity Horizon Scanning Programme.* London, UK: Department for Food, Environment and Rural Affairs.

Wade, P. R. 1995. Revised estimates of incidental kill of dolphins (Delphinidae) by the purse-seine tuna fishery in the eastern Tropical Pacific, 1959–1972. *Fishery Bulletin* 93: 345–354.

WAQIS. 1999. *Operations Manual.* Western Australian Quarantine and Inspection Service, Agriculture Western Australia: South Perth, Australia.

Webb, T. J., and D. Raffaelli. 2008. Conversations in conservation: Revealing and dealing with language differences in environmental conflicts. *Journal of Applied Ecology* 45: 1198–1204.

Weber, J.-L. 2011. Ecosystem services in green national accounting. In Koellner, T., ed., *Ecosystem Services and Global Trade of Natural Resources.* London, UK and New York, NY: Routledge.

Whitby, M. 2000. Challenges and options for the UK agri-environment: Presidential address. *Journal of Agricultural Economics* 51: 317–332.

Wilcox, C.1949. A charter for world trade. *University of Miami Law Review* 135.

Wilson, M. A., and J. P. Hoehn, 2006. Valuing environmental goods and services using benefit transfer: The state-of-the art and science. *Ecological Economics* 60: 335–342.

World Bank. 2017. GDP per capita, PPP (current international $). *International Comparison Program Database.* 1 July ed. Washington, DC: World Bank.

World Health Organization. 2010a. Biosecurity: An integrated approach to manage risk to human, animal and plant life and health. *International Food Safety Authorities Network (INFOSAN)*, Information Note No. 1/2010 — Biosecurity, 3 March 2010. Geneva, Switzerland: World Health Organization.

World Health Organization 2010b. *World Malaria Report 2010.* Geneva, Switzerland: World Health Organization.

World Health Organization. 2018. Yellow fever — Brazil. Disease outbreak news, 27 February [Online]. Geneva: World Health Organization. Available at: http://www.who.int/csr/don/27-february-2018-yellow-fever-brazil/en/ [Accessed 26th November 2018].

World Steel Association. 2018. *World Steel in Figures 2018.* Brussels, Belgium: World Steel Association.

World Trade Organization (WTO). 1998a. Panel Report, Australia — Measures affecting importation of salmon, Report of the Appellate Body, WT/DS18/AB/R. Geneva, Switzerland: World Trade Organization.

World Trade Organization (WTO). 1998b. Panel Report, Australia — Measures affecting importation of salmon, Report of the Panel, WT/DS18/R. Geneva, Switzerland: World Trade Organization.

World Trade Organization (WTO). 2007. Panel Report, Brazil — Measures affecting imports of retreaded tyres, WT/DS332/R. Geneva, Switzerland: World Trade Organization.

World Trade Organization (WTO). 2016. United States — Measures concerning the importation, marketing and sale of tuna and tuna products — Recourse to Article 22.6 by United States (DS381). Geneva, Switzerland: World Trade Organization.

World Trade Organization (WTO). 2018a. Available at: https://www.wto.org/english/thewto_e/whatis_e/tif_e/disp1_e.htm [Accessed 14th January 2018].

World Trade Organization (WTO). 2018b. General exceptions article. Available at: https://www.wto.org/english/docs_e/legal_e/gatt47_02_e.htm{\#}articleXX [Accessed 1st February 2018].

World Trade Organization (WTO). 2018c. SPS benefits. Available at: https://www.wto.org/english/tratop_e/sps_e/spsund_e.htm [Accessed 2nd February 2018].

World Trade Organization (WTO). 2018d. SPS explanation. Available at: https://www.wto.org/english/tratop_e/sps_e/spsagr_e.htm [Accessed 1st February 2018].

Worner, S. P., and M. Gevrey. 2006. Modelling global insect pest species assemblages to determine risk of invasion. *Journal of Applied Ecology* 43: 858–867.

Wynne, B. 1992. Uncertainty and environmental learning — Reconceiving science and policy in the preventive paradigm. *Global Environmental Change-Human and Policy Dimensions* 2: 111–127.

Yambrusic, E. S. 1987. *Treaty Interpretation: Theory and Reality.* Washington D.C. USA: University Press of America.

Yen, S. T., and C. L. Huang. 1996. Household demand for finfish: A generalized double-hurdle model. *Journal of Agricultural and Resource Economics* 21: 220–234.

Yoon, K. P., and C.-L. Hwang. 1995. *Multiple Attribute Decision Making: An Introduction.* London, UK: Sage Publications.

Zadocks, J., and R. Shein. 1979. *Epidemiology and Plant Disease Management.* Oxford, UK: Oxford University Press.

Zilberman, D. 2011. Economics of global trade and ecosystem services. In Koellner, T., ed., *Ecosystem Services and Global Trade of Natural Resources.* London, UK and New York, NY: Routledge.

Index

CPSIA information can be obtained
at www.ICGtesting.com
Printed in the USA
BVHW042254131019
560925BV00007B/9/P

9 781786 347770